Microscale Experiments for General Chemistry

Kenneth L. Williamson
Mount Holyoke College

John G. Little
Saint Mary's High School
Stockton, CA

Houghton Mifflin Company   Boston   New York

Senior Sponsoring Editor: Richard Stratton
Assistant Editor: Marianne Stepanian
Associate Project Editor: Christina Lillios
Senior Production/Design Coordinator: Jennifer Waddell
Manufacturing Coordinator: Marie Barnes
Associate Director of Marketing: Karen Natale
Editorial Assistant: Joy Park

Copyright © 1997 by Houghton Mifflin Company. All rights reserved.

No part of this work may be reproduced or transmitted in any form or by any means, electronic or mechanical, including photocopying and recording, or by any information storage or retrieval system without the prior written permission of Houghton Mifflin Company unless such copying is expressly permitted by federal copyright law. Address inquiries to College Permissions, Houghton Mifflin Company, 222 Berkeley Street, Boston, MA 02116-3764.

Printed in the U.S.A.

Library of Congress Catalog Card Number: 96-76978

Student Text ISBN: 0-669-41606-1
Exam Copy ISBN: 0-669-41858-7

123456789-BH-01 00 99 98 97

Table of Contents

Preface *v*

Introduction *xi*

Laboratory Safety and Waste Disposal *xxiii*

1. Observing a Chemical Reaction *1*
2. Densities of Organic Liquids *13*
3. Separation of a Mixture *23*
4. Crystallization *33*
5. Sublimation *41*
6. Solubility and Solutions *47*
7. Isolation of the Silver in a Dime *57*
8. Synthesis of Manganese (II) Chloride *65*
9. Analysis of Copper Oxide *69*
10. Gravimetric Determination of Chloride Ion Concentration *75*
11. Gravimetric Determination of Sulfate Ion Content *85*
12. Carbonate or Bicarbonate? *93*
13. Hydrates: Structure and Properties *99*
14. Periodic Trends in Physical Properties: Densities of Unknowns *105*
15. Distillation *113*
16. Avogadro's Law *125*
17. Charles' Law: Temperature/Volume Relationships for Gases *131*
18. Determination of the Molar Volume of a Gas and the Universal Gas Constant *137*
19. Oxidation-Reduction Reactions *145*
20. Analysis by Oxidation-Reduction Titration *157*

Table of Contents

21. Concentration of Hydrogen Peroxide   *165*
22. The Oxidation States of Nitrogen   *171*
23. Microelectrolysis of a Salt   *179*
24. Calorimetry: Heat of Fusion of Ice, Heat of Neutralization, Specific Heat of a Metal, and Heat of Solution   *187*
25. Thermodynamics of Oxidation of Acetone by Hypochlorite   *199*
26. Reaction Kinetics: The Aldol Condensation of Acetone with Benzaldehyde   *207*
27. Metal Reactivities   *219*
28. Nonmetal Reactivities   *227*
29. Synthesis and Quantitative Analysis of an Iron Compound   *233*
30. Partition Coefficient of an Organic Acid   *245*
31. Analysis of Commercial Vinegar   *253*
32. Gas Chemistry: Carbon Dioxide   *263*
33. Gas Chemistry: Ammonia   *271*
34. Gas Chemistry: Sulfur Dioxide   *279*
35. Determination of Molar Mass by Vapor Density and Freezing Point Depression   *287*
36. Solubility, Complex Ions, and Qualitative Analysis   *297*
37. Water Softening   *307*
38. Growing Crystals in Gels   *315*
39. Synthesis and Analysis of Aspirin   *329*
40. Synthesis of Esters   *337*
41. Oxidation-Reduction: Dyeing with Indigo, the Blue Jeans Dye   *343*
42. Synthesis of Slime   *351*

Appendix   Useful Data Tables   *363*

Preface

Microscale Experiments for General Chemistry is intended for use in the full-year course that is typically required of science and engineering majors. It uses very small-scale procedures to introduce first-year college students to experiments that exemplify the principles of general chemistry. Why should the traditional general chemistry laboratory be converted to microscale? Two words best summarize the answer: safety and cost. Experiments carried out on a scale 1/100th to 1/1000th the former scale are much safer for humans and the environment. Unlike traditional methods, microscale experiments expose students to miniscule amounts of materials that are toxic, corrosive, flammable, explosive, and carcinogenic. This benefits the environment as well since chemical waste is practically eliminated in microscale experiments.

Cost is the other reason for shifting to microscale experimentation. Since the experiments require such small amounts of chemicals, you buy less and have less waste to contend with. On average, it now costs at least five times more to dispose of chemicals than to purchase them. Organic chemicals can be burned in an incinerator, but heavy metals and metal ions must be buried in a secure landfill at great expense. We have paid particular attention to this problem by eliminating heavy metals from almost all experiments. When they are necessary, the quantity per student is often less than a milligram.

The Transition To Microscale

Some experiments, such as those dealing with the gas laws, will look familiar to you, while others will use unusual apparatus, such as a tiny filter flask fitted for sublimation. Most of the experiments in *Microscale Experiments for General Chemistry* can be performed with traditional apparatus—specialized apparatus simply makes the procedures easier. For example, we make frequent use of a 10 x 100 mm graduated glass tube we call a "reaction tube." A common 4-inch glass test tube can be used, but it is not graduated, nor long and narrow like a reaction tube. A gradual restocking and replacement program can be used to introduce new items into the standard set of apparatus.

In addition to specialized glassware, two items are highly recommended for microscale experimentation. The first is an *electric sandbath*. No more flames! The safest way to provide heat in the laboratory is the electric sandbath. Flammable organic liquids can be used in the laboratory with no worries. Students will not burn themselves nearly as often, and the sandbath provides a controlled, uniform source of heat. The second key piece of apparatus is a *top-loading balance*. A modern top-loading balance, sensitive to a milligram, is an essential item in the microscale laboratory.

vi Preface

Treating Laboratory Waste

Each experiment closes with a section titled "Cleaning Up," which gives students specific instructions regarding the treatment of all by-products from the experiment. Where possible, students learn how to convert potentially hazardous waste into material that can be flushed down the drain. In other instances, they learn how to reduce the volume of toxic materials, which drastically cuts the cost of disposal. Since students are dealing with the same waste problems that confront industry and government today, these treatment techniques teach them to look at the whole experiment, rather than strictly the part which produces the desired product, reaction, or result.

A Novel Approach

Investigations in the Laboratory

Science is concerned with finding answers to questions; as a result, every experiment involves an element of discovery, whether it be unknown substances to identify, testing the validity of commercial claims, or simply answering the question "What will happen when I mix these?" To the extent that it can be done safely, and in an environmentally responsible way, we have tried to build this element of investigation into every experiment.

A number of experiments are presented as unknowns to identify. Is that substance copper(I) or copper(II) oxide? Is that white material a carbonate or a bicarbonate? Can that black solid be purified by sublimation and identified? Can pure aspirin be isolated from a tablet? What is the chloride ion or the sulfate ion concentration in that unknown? How many waters of crystallization should be in that unknown hydrate? What percentage of the metal in a dime is real silver? Does that bottle of antiseptic really contain 3% hydrogen peroxide? How can a redox titration be used to identify an unknown—and unisolated—product?

Curriculum Flexibility

The experiments presented in this text are adaptable to a great variety of programs, both in terms of level of approach and the time needed to complete each assignment. Since one of us (JGL) teaches advanced placement chemistry in high school, many of these experiments were originally designed to fit the constraints of a fifty-minute laboratory period. This means that several experiments can, if necessary, be completed in an afternoon laboratory period. Part of this time-saving is a result of the great speed with which microscale experiments can be carried out. It simply does not require much time to boil a milliliter of water, or to set up a micro distillation apparatus that needs minimal clamping.

Authenticity

We have endeavored to illustrate chemical principles using apparatus that looks and feels like that used in the modern chemistry laboratory. For instance, some experiments involve the tools of a microbiologist, such as plastic pipettes and well plates, while others require the glass flasks, reaction tubes and distillation apparatus used by a chemist.

Innovative Techniques and Experiments

A number of fascinating, challenging, and informative experiments are included in this lab manual:

Separation of a Mixture. Students enjoy the challenge of separating a mixture of sand, iron, polyethylene, urea and naphthalene on a microscale. They make use of a tiny funnel and filter flask. The flask doubles as part of a sublimation apparatus for isolating naphthalene.

Sublimation. Simple apparatus introduces a technique that most students have never observed. Black, impure unknowns are converted to glistening crystals that can be identified by their melting points.

Solubility and Solutions. Solubility and solutions are explored on a microscale in a 24-well test plate and micro test tubes. Covalent and ionic substances are examined in a variety of solvents; a simple micro conductivity apparatus is used to test solutions.

Isolation of Silver from a Dime. You can't get silver from a modern dime, but a visit to a coin store will produce some 40-cent dimes, each of which provides enough silver for five students to run the experiment.

Redox Chemistry is thoroughly explored in a five-part experiment that has students exploring the interactions of nine oxidizing and reducing agents before venturing into vanadium redox chemistry. In another experiment, the iron content of an unknown is determined by permanganate titration.

Micro Electrolysis in a small U-tube is used to investigate copper(II) chloride, nickel nitrate, potassium iodide, zinc bromide, sodium chloride and table salt.

Calorimetry is thoroughly explored in the determination of the heat of fusion of ice, heat of neutralization, specific heat of a metal, heat of solution and heat of reaction (the oxidation of acetone with bleach). As an (optimal) alternative to the traditional thermometer, a digital thermometer or a digital probe connected to a computer adds a new dimension to these experiments.

Relative Activities of five metals and three non-metals (the common halogens) are determined on a microscale.

Spectrophotometric Analysis, Redox Titrations and Gravimetric Methods are used to analyze an unknown that the student synthesizes.

Gas Chemistry, on a microscale. Using easy-to-construct apparatus, experiments on carbon dioxide, ammonia and sulfur dioxide are conducted on a microscale. These unique experiments demonstrate the formation of highly-colored complexes, reactions with indicators, and precipitation of compounds from solution.

Freezing Point Depression and Vapor Density measurements are used to determine the molar masses of an unknown. The freezing point experiment may be easily interfaced to a computer, if desired.

Growing Crystals in Gels is a semester-long experiment that everyone enjoys watching. Liesegang rings and copper trees grow slowly in micro test tubes.

Sweet-Smelling Esters and Soap are prepared in an experiment that looks at both sides of the esterification/hydrolysis (saponification) equilibrium. The catalyst used in the microscale esterification reactions is a strong-acid ion exchange resin.

Aspirin is synthesized in a simple organic experiment. Pure aspirin is isolated from a tablet by crystallization from tert-butyl methyl ether in an experiment that demonstrates the utility of this important procedure. The new solvent tert-butyl methyl ether, a gasoline additive, is inexpensive, does not easily form peroxides and is not as flammable as diethyl ether.

Slime, a familiar toy, has much to teach us about the physical and chemical properties of a unique polymer. This borate ester is synthesized under a variety of conditions and then compared to the analogous silicate ester.

Computer Interfacing of Experiments

Don't discard those old computers! They may be useless for modern word processing and surfing the Internet, but they are great for collecting and displaying temperature, pressure, pH, absorbance, and conductance on a miniscale. The use of computers to log data is presented as an alternative to traditional methods with the full realization that not all institutions are fully equipped in this regard.

A site license for the software, a computer interface, and temperature, pressure, pH, and conductivity probes can be obtained for less than you would spend on one traditional melting point apparatus. We use these probes as optional data-gathering devices in the experiments on fractional distillation, the determination of molecular weight using melting point depression, and the calorimetric determination of heats of reaction, among others.

Instuctor's Guide

The Instructor's Guide is a vital part of this laboratory program and should be consulted for every experiment. It contains:

- Information for preparing all of the solutions used, warnings regarding unstable solutions and reagents, specific information regarding sources of supply, and detailed disposal and safety information, all of which will greatly facilitate setting up the labs.
- The quantities of reagents and apparatus for a laboratory section of twenty-four students.
- The placement of a particular experiment in the curriculum is detailed along with effective pairings of short experiments.
- The answers to all prelaboratory and postlaboratory questions, as well as sample data and sample calculations, are given for each experiment.
- The time needed to set up the experiments, the degree of difficulty and time needed to complete the experiments, information particularly valuable for those using this text in advanced placement programs.

In short, we have tried to give the instructor the benefits of our half-century of combined experience in teaching laboratory chemistry, over twenty years of which has been in teaching microscale experiments.

Acknowledgments

We would like to acknowledge the encouragement and patience of the editors at Houghton Mifflin Company, in particular Richard Stratton, Marianne Stepanian, and Christina Lillios. We also wish to express our thanks to the following reviewers who made so many useful comments and suggestions:

R. K. Bridwhistell
University of West Florida

Nathan J. Cook
Colgate University

Alfred J. Lata
The University of Kansas

C. Dean Mitchell
California State University, Fresno

Maureen A. Scharberg
San Jose State University

Susan M. Shih
College of DuPage

Susan Thornton
Montgomery College

Nancy S. True
University of California, Davis

Introduction

Welcome to the chemistry laboratory! This laboratory manual presents a new approach for carrying out experiments—on a microscale.

For reasons primarily of safety and cost, a trend is emerging toward carrying out work in the laboratory on a microscale, a scale one-tenth to one-thousandth that previously used. The use of smaller quantities of chemicals exposes the laboratory worker to smaller amounts of toxic, flammable, explosive, carcinogenic, and teratogenic material. Microscale experiments can be carried out very quickly and the cost of chemicals is, of course, greatly reduced.

As will be seen, some of the equipment and techniques are different from that you might have seen in previous chemistry laboratories. For that reason we present here an introduction to the equipment and some of the techniques.

Equipment Used in Separations

The Dutch word for chemistry is *scheikunde,* which translates as "the art of separation." Separation and purification of the components of mixtures is the first and most important step in chemical investigations. You will carry out many separations and purifications in this course. Liquids with different boiling points can be separated by distillation in a *one-piece distillation apparatus* (A), some solids can be sublimed away from nonvolatile impurities in a *sublimation apparatus* (B), and soluble substances can be separated from insoluble ones using *filtration apparatus*, consisting of a *Hirsch funnel* with *filter disk*, which is mounted in a 25-mL *filter flask* (C). The flask is the same one used in the sublimation apparatus.

Supporting Apparatus

To hold apparatus during filtration, distillation, or a number of other operations, you will need a *ringstand* (D) to which are attached *clamp holders* (E) and *clamps* (F).

Measuring Apparatus

Chemistry is a quantitative science. You will usually need to know how much material you are working with. The *top-loading balance* (G), used to weigh substances to ±0.001 g, is the most important piece of apparatus you will use. Liquids are measured in *1-mL pipettes* (H and I) that can be filled using a *syringe* (H) or a *pipette pump* (I). **Never use your mouth!** Larger quantities of liquids can be measured to ±1 mL in a *50-mL graduated cylinder* (J) or to ±0.02 mL in a *5-mL buret* (K). We will also measure liquids by large drops from a *Pasteur pipette* with a *rubber pipette bulb* (L), or in small drops from a polyethylene *microtip Beral pipette* (M).

Temperature is, of course, measured with a *thermometer* (N). If your thermometer is filled with mercury, be warned that mercury is very toxic. Report a broken mercury thermometer to your instructor immediately.

> **Safety Information** **Caution!** Notify your instructor immediately if you break a thermometer. Mercury is very toxic.

Containers

A wide variety of containers are used in the laboratory. Some, such as *capped vials* (O), *shell vials* (P), or *polyethylene vials* (Q), are used to store chemicals.

Check In

Your first duty will be to check in to your assigned desk. The identity of much of the apparatus should already be apparent from the preceding outline.

Check to see that your thermometer reads about 22 to 25°C (20°C = 68°F), normal room temperature. Examine the mercury column to see if the thread is broken. When intact, the mercury is a continuous column from the bulb up.

Washing and Drying Laboratory Equipment

> **Safety Information** Clean apparatus immediately.

Considerable time can be saved by cleaning each piece of equipment soon after use, for you will know at that point what contaminant is present and be able to select the proper method for removal. A residue is easier to remove before it has dried and hardened.

Glass Tubes, Thermometers, and Rubber Stoppers

Insertion of a glass tube into a rubber connector, adapter, or hose is easy if the glass is lubricated with a very small drop of glycerol. Grasp the tube very close to the end to be inserted; if it is grasped at a distance, especially at a bend, the pressure applied for insertion may break the tube and result in a serious cut.

If a glass tube or thermometer should become stuck to a rubber connector, it can be removed by painting on glycerol and forcing the pointed tip of an 18-cm spatula between the rubber and glass. Another method is to select a cork borer that fits snugly over the glass tube, moisten it with glycerol, and slowly work it through the connector. When the stuck object is valuable, such as a thermometer, the best policy is to cut the rubber with a sharp knife.

Weighing and Measuring

Weighing and Measuring

To carry out a reaction, known quantities of reactants and solvents are usually heated together for a period of time varying from a few seconds to several hours. Some substances must be measured more accurately than others; time can be saved by knowing which need to be measured only approximately.

The Top-Loading Electronic Balance

The single-pan electronic balance is capable of weighing to ±0.001 g and has a capacity of about 120 g. It is the single most important piece of apparatus in the microscale laboratory. In this text most of the quantitative measurements will use the balance. One can easily weigh a liquid to ±0.001 g but it is not nearly so easy to measure to ±1 microliter (0.001 g of a liquid having density 1.00) using volumetric devices.

On these top-loading balances it is most convenient to weigh at least one of the reactants (solid or liquid) directly into the reaction container. The flask, reaction tube, or vial is placed on the pan of the balance, the tare lever or button is pressed so that the balance reads zero, and then the substance to be weighed is added. If a procedure calls for, say, 40 mg of a substance it is usually acceptable to weigh ±3 mg, *as long as the amount actually weighed is recorded*. One can waste a lot of time trying to weigh a substance to the nearest milligram.

Transfer of a Solid

It is often convenient to weigh reagents on glossy weighing paper and then transfer the chemical to the reaction container. The success of an experiment often depends on using just the right amount of certain reagents. Inexperienced workers might think that if one-tenth of a milliliter of a reagent will do the job, then two-tenths of a milliliter will do the job twice as well. Such assumptions are usually erroneous.

Measuring Liquids

Liquids can be measured by either volume or mass according to the relationship

$$\text{Volume (mL)} = \frac{\text{mass (g)}}{\text{density (g/mL)}}$$

> **Safety Information** Never pipette by mouth.

Modern Erlenmeyer flasks and beakers have approximate volume calibrations fused into the glass, but these are *very* approximate. The graduations on the reaction tube are accurate enough for most purposes. Somewhat more accurate volumetric measurements are made in 10-mL graduated cylinders, For volumes less than about 4 mL, use a graduated pipette. **Never** apply suction to a pipette by mouth. The pipette can be fitted with a small rubber bulb, or better, a pipette pump. A Pasteur pipette can be converted into a calibrated pipette with the addition of a plastic syringe body or you can calibrate it at 0.5, 1.0, and 1.5 mL and put three file scratches on the tube; this eliminates the need to use a syringe with this Pasteur pipette in the future. You will find among your equipment a 1-mL pipette, calibrated in hundredths of a milliliter. Determine whether it is designed to *deliver* 1 mL or to *contain* 1 mL between the top and bottom calibration marks. For our purposes, the latter is the better pipette.

Tares

Tare = mass of empty container

The tare of a container is its mass when empty. Throughout this laboratory course it will be necessary to know the tares of containers so that the masses of the compounds within can be calculated. If identifying marks can be placed on the containers (e.g., with a diamond stylus) you may want to record tares for frequently used containers in your laboratory notebook.

The Electrically Heated Sand Bath

An electrically heated sand bath has several advantages. The total surface area emitting heat is small, so it is possible to have the hands very near the heat source without burning them. The relatively poor heat conduction of sand results in a very large temperature difference between the top of the sand and the bottom. Thus, depending on the immersion depth in the sand, a similarly wide temperature range will be found in the container being heated, a very useful feature of this heating method. Also, since the surface area of the sand bath is small, the air above the heater is cool, making the use of air condensers in distillation very efficient.

Prevention of Superheating

Regardless of the scale of reaction, the container, or the means of heating it, it is possible for liquids to *superheat*, that is, to remain in a quiescent state above the boiling point. When boiling does begin it can do so with explosive violence, expelling the liquid out of the container. In a somewhat milder form this is called *bumping*. To prevent it add a *boiling stone*, a *boiling chip*, or a *boiling stick*.

As a boiling stone or stick warms up in the solution, tiny bubbles of air escape from the surface and form the nuclei on which bubbles of vapor form to give even boiling. Never add a boiling chip or stick, charcoal, or any other solid to a solution suspected of being above its boiling point since violent boiling will be initiated. Rapid stirring will often alleviate superheating.

Laboratory Equipment

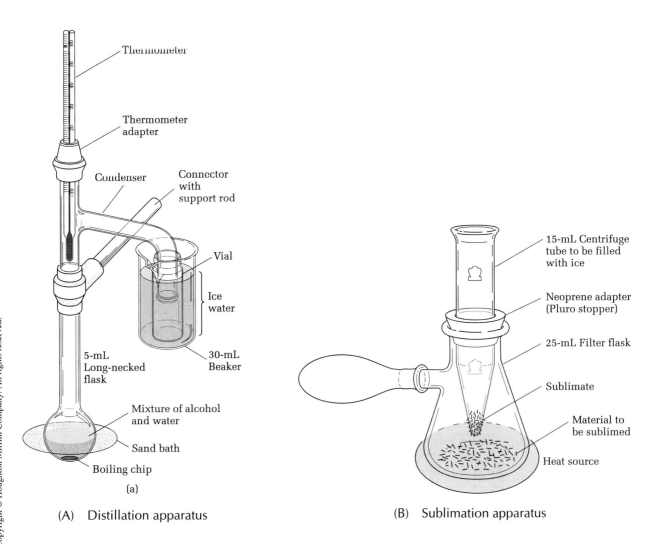

(A) Distillation apparatus

(B) Sublimation apparatus

xvi   Introduction

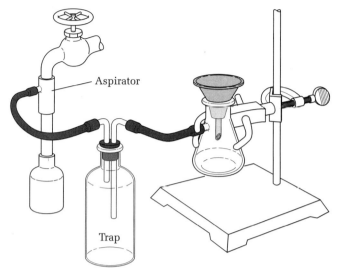

(C) Filtration assembly

(E) Clamp holder

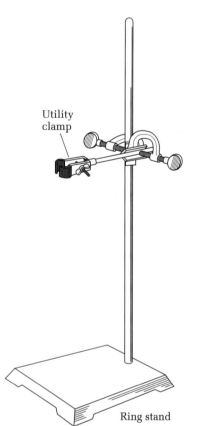

(D) Ring stand

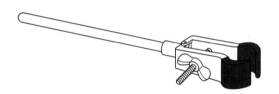

(F) Clamp

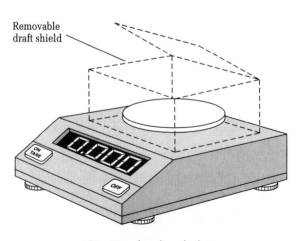

(G) Top-loading balance

Introduction xvii

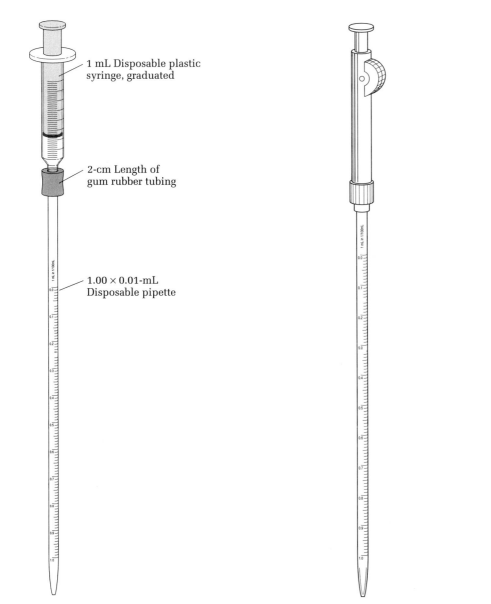

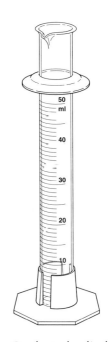

(H) Graduated pipette delivery system

- 1 mL Disposable plastic syringe, graduated
- 2-cm Length of gum rubber tubing
- 1.00 × 0.01-mL Disposable pipette

(I) Graduated disposable pipette and pipette pump

(J) Graduated cylinder

(K) Burette

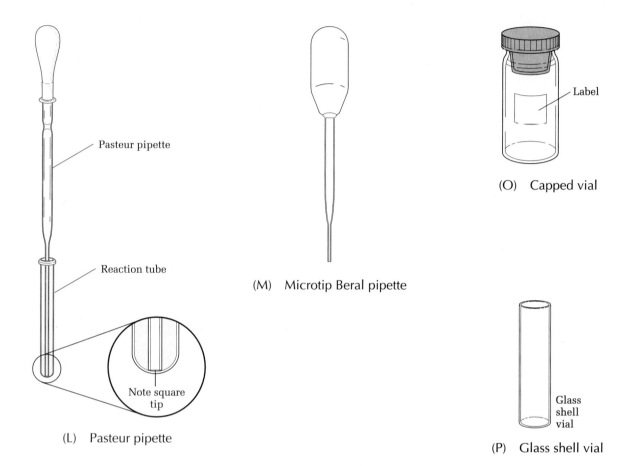

(L) Pasteur pipette

(M) Microtip Beral pipette

(O) Capped vial

(P) Glass shell vial

(N) Thermometer

(Q) Polyethylene vial

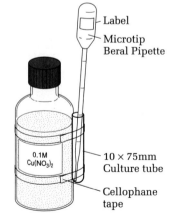

(R) Reagent bottle with graduated Beral pipette

Introduction xix

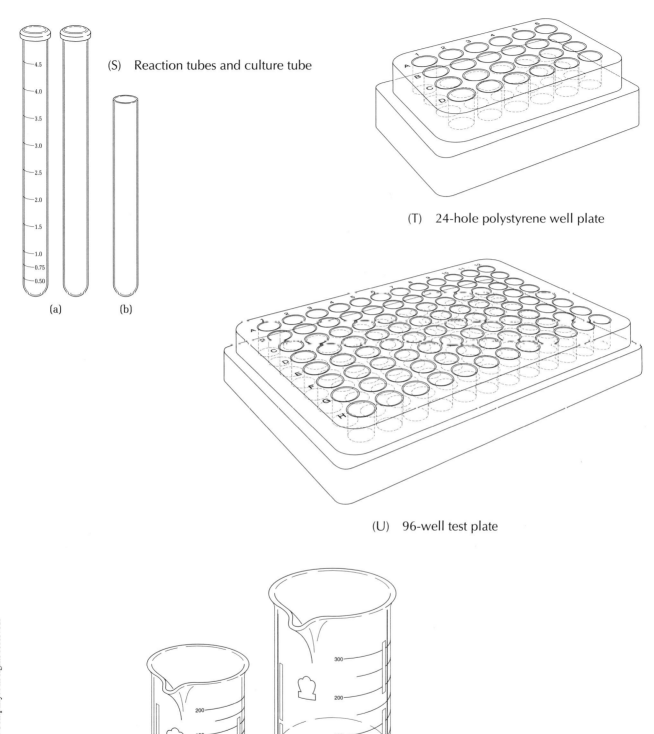

(S) Reaction tubes and culture tube

(T) 24-hole polystyrene well plate

(U) 96-well test plate

(V) Beakers

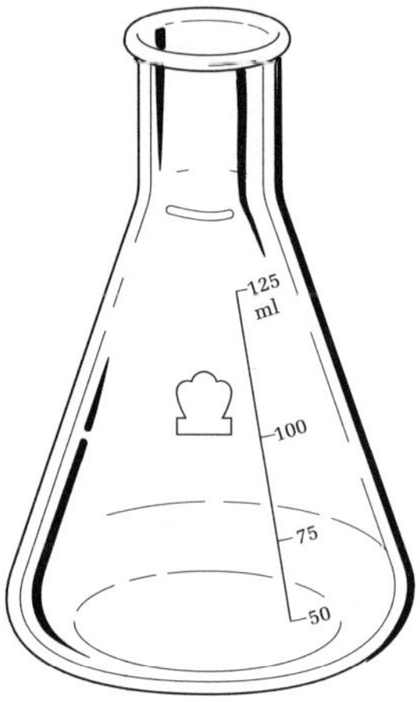

xx   Introduction

(W)   Erlenmeyer flask

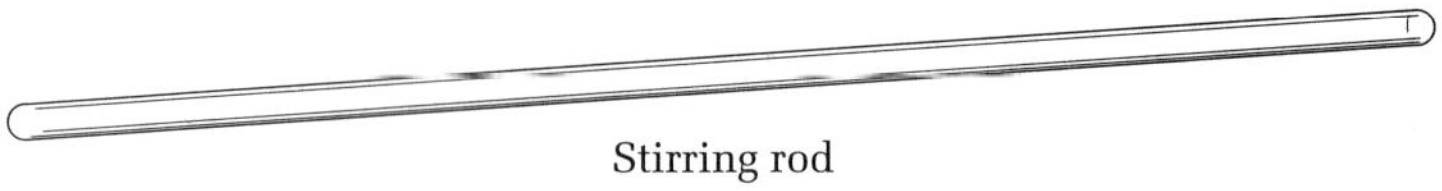

(X)   Stirring rod

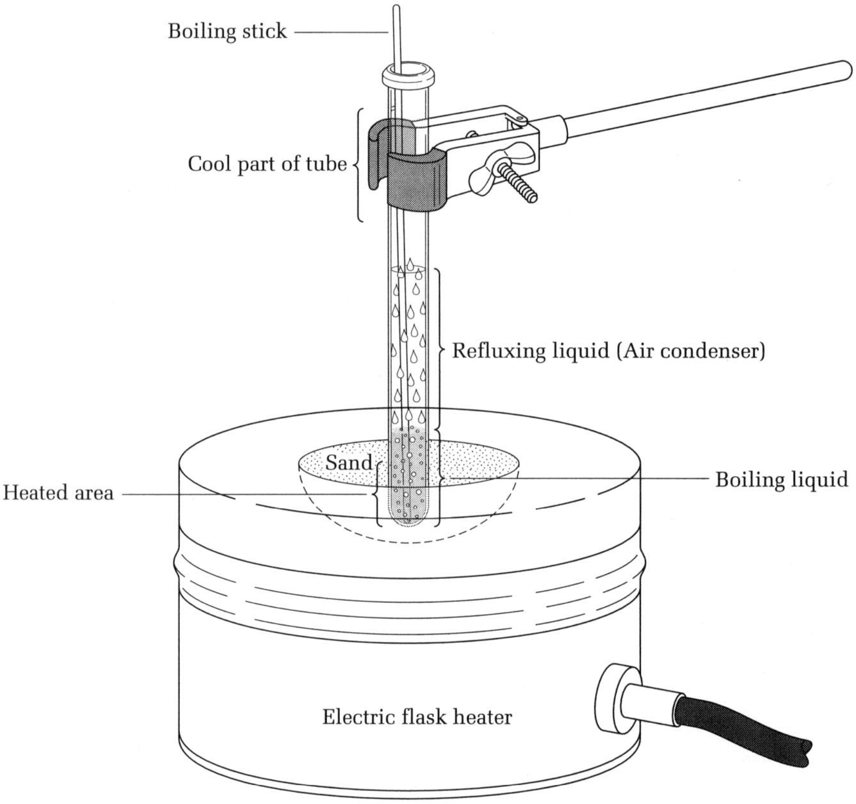

(Y)   Electrically-heated sand bath

(Z)   Polyethylene wash bottle

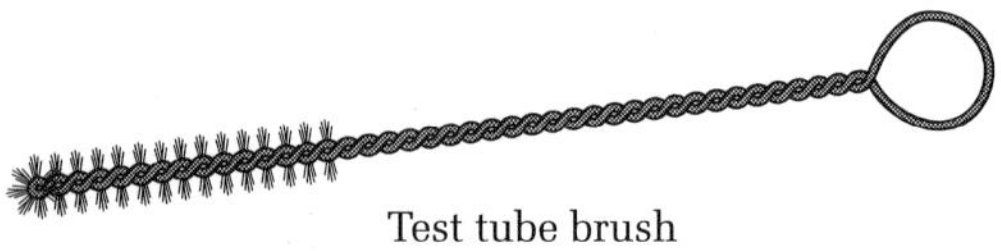

(AA) Test tube brush

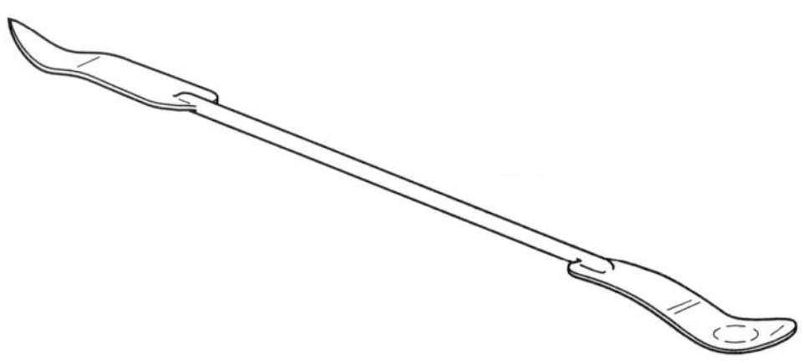

(BB) Spatula

Introduction   xxi

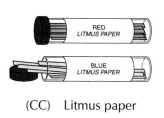

(CC) Litmus paper

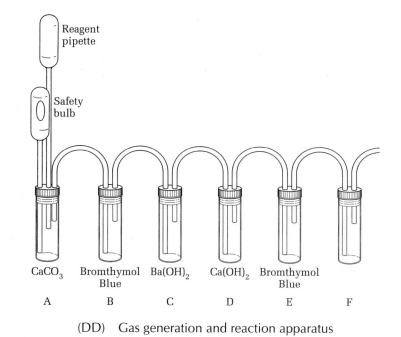

(DD) Gas generation and reaction apparatus

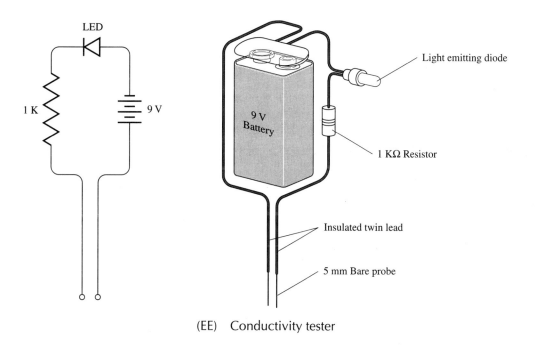

(EE) Conductivity tester

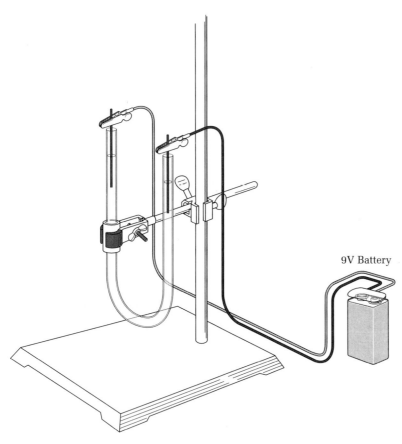

(FF) Electrolysis U-tube

(GG) Safety goggles

(HH) Solvent safety can

(II) Fire extinguisher

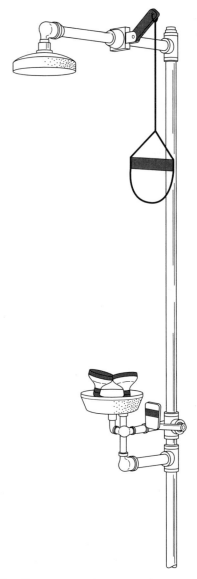

(JJ) Emergency shower and eye wash

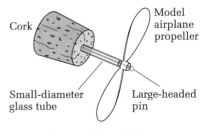

(KK) Airflow indicator

Laboratory Safety and Waste Disposal

Microscale chemical experiments are much safer than their traditional counterparts, which are usually run on a scale 10 to 1,000 times larger. Regardless of the scale, however, the chemistry laboratory is an excellent place to learn and practice safety. You will find that the common sense procedures practiced here also apply to other laboratories and to the shop, kitchen, and studio.

The safety information outlined here is not comprehensive. Throughout this text you will find specific safety information and cautions presented in each experiment. For a relatively brief and more thorough discussion of all the topics in this chapter, you should read the first 35 pages of *Safety in Academic Chemistry Laboratories* (American Chemical Society, Washington, DC, 1990). The best book on the subject is *Prudent Practices in the Laboratory, Handling and Disposal of Chemicals* (National Research Council, Washington, DC, 1995). One of us (KLW) was one of the coauthors of an earlier *Prudent Practices*.

General Rules

Know the safety rules of your particular laboratory. Know the locations of emergency eyewash stations, fire extinguishers, and safety showers, and how to use them. Never eat, drink, or smoke in the laboratory. Don't work alone. Perform no unauthorized experiments, and don't distract your fellow workers; horseplay has no place in the laboratory.

Eye protection is extremely important. Safety glasses of some type must be worn at all times. Contact lenses must never be worn. Reagents can get under a contact lens and cause damage to the eye before the lens can be removed. It is very difficult to remove a contact lens from the eye after a chemical splash has occurred. Some organic liquids or their vapors can dissolve in soft contact lenses.

Ordinary prescription eyeglasses do not offer adequate protection. *Laboratory safety glasses* (GG) should be of plastic or tempered glass. If you do not have such glasses, wear goggles that afford protection from splashes and objects coming from the side as well as the front. If plastic safety glasses are permitted in your laboratory, they should have side shields.

Dress sensibly in the laboratory. Wear shoes, not sandals or cloth-top sneakers. Confine long hair and loose clothes. Don't wear shorts. Never use mouth suction to fill a pipette, and wash your hands before leaving the laboratory. Don't use a solvent to remove chemicals from the skin. This will only hasten the absorption of the chemical through the skin.

Working with Flammable Substances

You will be working with several flammable organic liquids in the course of carrying out these experiments, although never in quantities of more than a few milliliters.

Bulk organic solvents should be dispensed from *solvent safety cans* (HH), not large glass bottles. Flames are not intended to be used at any time in the course of carrying out the experiments in this text, so the danger of fire is greatly reduced. A fire in a beaker can be smothered by covering it with a flat object, even your laboratory notebook. A somewhat larger fire will require a fire extinguisher directed at the base of the fire.

Should a person's clothing catch fire and a safety shower is close at hand, shove the person under it. Otherwise, shove the person down and roll him or her over to extinguish the flames. It is extremely important to prevent the victim from breathing the hot vapors that rise past the mouth. The safety shower might then be used to extinguish glowing cloth that is no longer aflame. The so-called fire blanket should not be used—it tends to funnel flames past the victim's mouth, and clothing continues to char beneath it. However, it is useful for retaining warmth to ward off shock after the flames are out.

Because organic solvents are used in the course of these experiments, the laboratory should be equipped with a carbon dioxide or dry chemical (monoammonium phosphate) *fire extinguisher* (II). To use this type of extinguisher lift it from its support, pull the ring to break the seal, raise the horn, aim it at the base of the fire, and squeeze the handle. Do not hold onto the horn because it will become extremely cold. Do not replace the extinguisher; report the incident so the extinguisher can be refilled.

Dispose of chemicals only in the manner specified in the **Cleaning Up** sections of this book. When disposing of certain chemicals, be alert to the possibility of spontaneous combustion. This may occur with strong oxidizing agents such as nitric acid, permanganate ion, and peroxides; alkali metals such as sodium; or very finely divided metals such as zinc dust. These materials can set fire to filter paper in waste containers.

Working with Corrosive Substances

Handle strong acids, alkalis, dehydrating agents, and oxidizing agents carefully so as to avoid contact with the skin and eyes and to avoid breathing the corrosive vapors that attack the respiratory tract. All strong concentrated acids attack the skin and eyes. Concentrated sulfuric acid is both a dehydrating agent and a strong acid that will cause very severe burns. Nitric acid also causes bad burns.

Sodium, potassium, and ammonium hydroxides are common strong bases you will encounter. The first two are extremely damaging to the eye, and ammonium hydroxide is a severe bronchial irritant. Like sulfuric acid, sodium hydroxide and calcium oxide are powerful dehydrating agents. Their great affinity for water will cause burns to the skin. Because they release a great deal of heat when they react with water, to avoid spattering, they should always be added to water rather than water being added to them. That is, the more dense substance should always be added to the less dense one so that rapid mixing results as a consequence of the law of gravity.

Should one of these substances get on the skin or in the eyes, wash the affected area with very large quantities of water using the *safety shower and/or eyewash fountain* (JJ). Do not attempt to neutralize the reagent chemically. Remove contaminated clothing so that thorough washing can take place. Take care to wash the reagent from under the fingernails.

When you are using very small quantities of these reagents, no particular safety equipment is needed except safety glasses. Take care not to let the reagents, such as sulfuric acid, run down the outside of a bottle or flask and come in contact with the fingers. Wipe up spills immediately with a very damp sponge, especially in the area around the balances. Pellets of sodium and potassium hydroxide are very hygroscopic and will dissolve in the water they pick up from the air; therefore, they should be wiped up very quickly. When working with larger quantities of these corrosive chemicals, wear protective gloves; with still larger quantities, use a face mask, gloves, and a neoprene apron. The corrosive vapors can be avoided by carrying out work in a good exhaust hood.

Working with Toxic Substances

It is impossible to avoid handling every known or suspected toxic substance, so it is wise to know what measures should be taken to protect yourself from them. Because the eating of food or the consumption of beverages in the laboratory is strictly forbidden, and because one should never taste material in the laboratory, the possibility of poisoning by mouth is remote. Be more careful than your predecessors—the hallucinogenic properties of LSD and *all* artificial sweeteners were discovered by accident. The two most important measures to be taken, then, are avoiding skin contact by wearing the proper type of protective gloves (see next section) and avoiding inhalation by working in a good exhaust hood.

Many of the chemicals used in this course will not be familiar to you. Their properties can be looked up in reference books, a very useful one being the *Aldrich Catalog Handbook of Fine Chemicals* (Aldrich Chemical Co., 1001 West Saint Paul Ave., Milwaukee, WI 53233). It is interesting to note that 1,4-dichlorobenzene is listed as a "toxic irritant" and naphthalene is listed as an "irritant." Both are used as mothballs. Camphor, used in vaporizers, is classified as a "flammable solid irritant." Salicylic acid, which we will use to synthesize aspirin (Experiment 39), is listed as "moisture-sensitive toxic." Aspirin (acetylsalicylic acid) is classified as an "irritant." To put things in some perspective, nicotine is classified as "highly toxic."

Consult M. A. Armour, L. M. Browne, and G. L. Weir, *Hazardous Chemicals Information and Disposal Guide* (Department of Chemistry, University of Alberta, Edmonton, Alberta, Canada T6G 2G2, 3rd ed., 1988) for information on truly hazardous chemicals.

Because you may not have had previous experience working with chemicals, the experiments you will carry out in this course do not involve the use of known carcinogens, although you will work routinely with flammable, corrosive, and toxic substances.

Gloves

Be aware that "protective gloves" in the laboratory may not offer much protection. Polyethylene and latex rubber gloves are very permeable to many organic liquids. An undetected pinhole can mean long-term contact with reagents. Disposable polyvinyl chloride (PVC) gloves offer reasonable protection from contact with aqueous solutions of acids, bases, and dyes, but no one type of glove is useful as protection against all reagents. It is for this reason that no less than 25 different types of chemically resistant gloves are available from laboratory supply houses.

It is probably safer not to wear gloves and immediately wash your hands with soap and water after accidental contact with any harmful reagent or solvent than to wear inappropriate or defective gloves.

Using the Laboratory Hood

Modern practice dictates that in laboratories where workers spend most of their time working with chemicals, there should be one exhaust hood for every two people. This precaution, however, is often not possible in the beginning chemistry laboratory. Because you will be carrying out experiments on a microscale, many operations formerly carried out in the hood can now be carried out at the desk because the concentration of vapors will be minimal.

The hood offers a number of advantages for work with toxic and flammable substances. Not only does it draw off the toxic and flammable fumes, it also affords an excellent physical barrier on all four sides of a reacting system when the sash is pulled down. And should a chemical spill occur, it is nicely contained within the hood.

It is your responsibility each time you use a hood to see that it is working properly. You should find some type of indicating device that will give you this information on the hood itself. A simple propeller on a cork works well as an *airflow indicator* (KK). The hood is a backup device. Don't use it alone to dispose of chemicals by evaporation. Toxic and flammable fumes should be kept trapped or condensed in some way and disposed of in the prescribed manner. Except when you are actually carrying out manipulations on the experimental apparatus, the sash of the hood should be pulled down. The water, gas, and electrical controls should be on the outside of the hood so that it is not necessary to open the hood to adjust them. The ability of the hood to remove vapors is greatly enhanced if the apparatus is kept as close to the back of the hood as possible. Everything should be at least 15 cm back from the hood sash. Chemicals should not be stored permanently in the hood but should be removed to ventilated storage areas. If the hood is cluttered with chemicals, you will not have good, smooth airflow and adequate room for experiments.

Waste Disposal: Cleaning Up

Spilled solids should simply be swept up and placed in the appropriate solid waste container. This should be done promptly because many solids are hygroscopic and become difficult, if not impossible, to sweep up in a short time. This is particularly true of sodium hydroxide and potassium hydroxide.

Spilled acids should be neutralized. Use sodium carbonate or, for larger spills, cement or limestone. For bases, use sodium bisulfate. If the spilled material is very volatile, clear the area and let it evaporate, provided there is no chance of igniting flammable vapors. Other liquids can be taken up into such absorbents as vermiculite, diatomaceous earth, dry sand, or paper towels. Be particularly careful when wiping up spills with paper towels. If a strong oxidizer is present, the towels can later ignite. Unless you are sure the spilled liquid is not toxic, wear gloves when using paper towels or a sponge to remove the liquid.

Cleaning Up

In the not too distant past it was common practice to wash all unwanted liquids down the drain and to place all solid waste in the trash basket. Never a wise practice, for environmental reasons, this is no longer allowed by law.

Some reactions utilize a flammable solvent and often involve the use of a strong acid, strong base, an oxidant, reductant, or a catalyst. None of these should be washed down the drain or placed in the wastebasket. Waste should be placed in the proper container. Broken glass should be placed in its own special receptacle (to protect laboratory cleaning personnel).

Nonhazardous waste encompasses such solids as paper, corks, sand, alumina, and sodium sulfate. These ultimately will end up in a sanitary landfill (the local dump). Any chemicals that are leached by rainwater from this landfill must not be harmful to the environment.

Nonhalogenated solid organic waste will go to an incinerator where it will be burned. A container should be supplied for various compatible hazardous wastes (usually solids). Some hazardous wastes are incompatible (oxidants with reductants, for example), and so you may find several different containers for these in the laboratory, as well as a container for heavy metals.

To dispose of small quantities of a hazardous waste, for example, concentrated sulfuric acid, the material must be carefully packed in bottles and placed in a 55-gal drum called a lab pack, to which is added an inert packing material. The lab pack is carefully documented and then hauled off by a bonded, licensed, and heavily regulated waste disposal company to a site where such waste is disposed. Formerly, many hazardous wastes were disposed of by burial in a "secure landfill." The kinds of hazardous waste that can be thus disposed of have become extremely limited in recent years, and much of it undergoes various kinds of treatments at the disposal site (for example, neutralization, incineration, reduction) to put it in a form that can be safely buried in a secure landfill or flushed to a sewer. Relatively few places exist for approved disposal of hazardous waste. For example, in the United States there are none in New England, so most hazardous waste from this area is trucked to South Carolina, more than 1,000 km away! The charge to small generators of waste is usually based on the volume of waste. Therefore, 1000 mL of a 2% cyanide solution would cost much more to dispose of than 20 g of solid cyanide, even though the total amount of this poisonous substance is the same. It now costs much more to dispose of most hazardous chemicals than it does to purchase them new.

American law states that a material is not a waste until the laboratory worker declares it a waste. For pedagogical and practical reasons, therefore, we would like you to regard the chemical treatment of the by-products of each reaction in this text as part of the experiment.

In the section entitled "Cleaning Up" at the end of each experiment, the goal is to reduce the volume of hazardous waste, to convert hazardous waste to less hazardous waste, or to convert it to nonhazardous waste. The simplest example is concentrated sulfuric acid. As a by-product from a reaction, it is obviously hazardous. But after careful dilution with water and neutralization with sodium carbonate, the sulfuric acid becomes a dilute solution of sodium sulfate, which in almost every locale can be flushed down the drain with a large excess of water. Anything flushed down the drain must be accompanied by a large excess of water. Dilute solutions of heavy metal ions can be precipitated as their insoluble sulfides or hydroxides. The precipitate may still be a hazardous waste, but it will have a much smaller volume.

Our goal in "Cleaning Up" is to make you more aware of all aspects of an experiment. Waste disposal is now an extremely important aspect. Check to be sure the procedure you use is legal in your location. Three sources of information have been used as the basis of the procedures at the end of each experiment: the *Aldrich Catalog Handbook of Fine Chemicals,* which gives brief disposal procedures for every chemical in their catalog; *Prudent Practices for the Disposal of Chemicals from Laboratories* (National Academy Press, Washington DC, 1995); and *Hazardous Chemicals Information and Disposal Guide.* The last should be on the bookshelf of every laboratory. This 300-page book gives detailed information about 287 hazardous substances, including physical properties, hazardous reactions, physiological properties and health hazards, spillage disposal, and waste disposal. Many of the treatment procedures in the Cleaning Up sections are adaptations of these procedures. *Destruction of Hazardous Chemicals in the Laboratory*, by G. Lunn and E. B. Sansone (Wiley, New York, 1990), complements this book.

The area of waste disposal is changing rapidly. Many different laws apply—local, state, and federal. What may be permissible to wash down the drain or evaporate in the hood in one jurisdiction may be illegal in another, so before carrying out this part of the experiment, check with your college or university waste disposal officer.

QUESTIONS

1. Define
   a. Hygroscopic

   b. Volatile

   c. Carcinogen

   d. Corrosive

   e. Caustic

   f. Heavy metal

2. List three possible concerns when selecting gloves for laboratory use.

3. One should never store piles of oily rags in a work area. Why not? What similar situation is mentioned in the preceding sections?

EXPERIMENT 1

Observing a Chemical Reaction

Introduction

The scientific method involves, in part, making careful observations from which conclusions can be drawn. It is important at the outset to distinguish between observation and conclusion. For instance, if a chemical reaction evolves a gas it would be correct to say, based on observation, that a gas is formed, but it would be incorrect to say that carbon dioxide gas was evolved without testing the gas in some way.

Observations can be made by sight, feel, smell, and occasionally, under special circumstances, by taste and sound. Instruments can help us make observations, especially quantitative observations. A reaction may feel warm; a thermometer will measure just how warm.

You should make careful notes when carrying out an experiment. An observation that might not seem important at the time may later turn out to be important, so write it down. *Quantitative* observations, observations that involve a measurement, are valuable in reaching conclusions about an experiment, so whenever possible make sure your measurements are as accurate as possible. It is better to record that a flask contains 55 mL of a liquid rather than to simply state that it is "about half full."

As you will see, the laws of science in general and chemistry in particular do not change from day to day or week to week. If two people perform the same chemical experiment in exactly the same way, they will get exactly the same results, time after time. The results of a good experiment are therefore said to be *reproducible*.

Procedure Summary

Your objective in the present experiment is to make careful observations of a chemical reaction and to make, wherever possible, quantitative measurements. In the process, you will distinguish between observations and conclusions, and you will attempt to determine whether certain of your observations are reproducible. You will be observing the behavior of the types of reagents that you will encounter throughout the year; many of the phenomena you see here will appear again in later experiments.

Carefully note in these experiments the distinction between what you observe and what you can legitimately conclude. Do any of these experiments tell you anything about the conservation of mass in a chemical reaction?

Experiment 1 Observing a Chemical Reaction

Prelaboratory Assignment

Read the Introduction and Procedure sections and answer the Prelaboratory Questions on the Report Sheet before you come to the laboratory.

Materials

Apparatus

Safety goggles
96–well test plate
1-mL graduated pipette
Glass rod or plastic toothpick
Thermometer
3 reaction tubes or 10×75-mm culture tubes
100-mL beaker
Small ringstand
Small three-prong clamp
50-mL beaker to hold flask and culture tubes for weighing
Magnifying glass (optional)
10-mL Erlenmeyer flask
Wax pencil (if unmarked tubes are used)
Paper towels to dry steel wool
1-mL graduated Beral pipette

Reagents

3 M hydrochloric acid
3 M acetic acid
Steel wool (grade 0000)
1.5 M sodium bicarbonate solution
1 M calcium chloride solution
3 M sulfuric acid solution
1 M sodium carbonate solution

Safety Information

1. **Safety goggles must be worn at all times in the laboratory.**
2. **Handle mercury-containing thermometers very carefully.** Don't let the thermometer roll off the desk or put it into a container that can tip over. Don't use it as a stirring rod. If it should accidentally become broken, notify your instructor immediately because mercury vapor is very toxic; the liquid must be properly cleaned up and disposed of.
3. **Avoid contact with acids.** Wipe up any spills immediately and wash your hands.
4. **Use care when smelling chemicals.** Don't sniff the bottle directly; waft the odor toward your nose with your hand. See Figure 1.1.

Figure 1.1 Be careful when checking odors. Smell a substance by wafting its vapors gently toward your nose.

Procedure

1. Using the plastic dropper that comes with the reagent bottle (Figure 1.2), half fill a well in the corner of a 96–well test plate (Figure 1.3) with 3 M hydrochloric acid. The designation "3 M" (three molar, three moles per liter) refers to the concentration of the reagent. As far away on the plate as possible half fill another well with 3 M acetic acid. Do not mix up the droppers. Carefully note the odors of the two acids and their appearances. To each acid add a small, loose ball of fine steel wool, about 3 mm in diameter (there is a ruler in

the Appendix of this book). Push it into the acid with a glass stirring rod. Note evidence of any reaction. Again carefully note the odors and then set the well plate aside.

2. Push a loosely packed 10-mm ball of steel wool halfway down a reaction tube or a 10 × 75-mm culture tube (Figure 1.4). *Without delay,* add to the steel wool 3 drops of 3 *M*

Figure 1.2 Reagent bottle with 1-mL graduated Beral pipette.

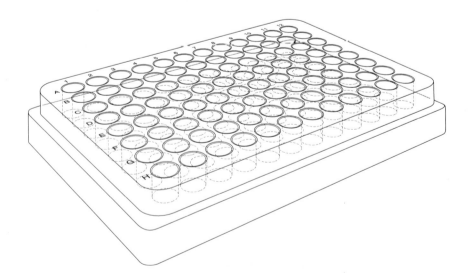

Figure 1.3 A 96–well test plate.

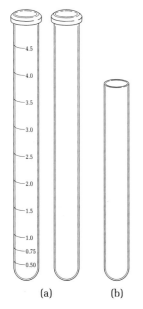

Figure 1.4 (a) Reaction tube, 10 × 100 mm, graduated and ungraduated and (b) a 10 × 75-mm culture tube.

acetic acid. Clamp the tube in an inverted position with the mouth of the tube slightly below the surface of the water in a 100-mL beaker. Note the level of the water in the tube (Figure 1.5). When no further change is noted in the level of the water inside the tube, adjust the tube so the water levels inside and outside the tube are the same and estimate as carefully as possible the level of the water in the tube using the graduations marked on it.

Note that the Data and Observations section of the Report Sheet calls for volume readings. If you are using an unmarked tube, mark the initial water level inside the tube and record the length of the air column above the water line, inside the tube. A wax pencil can be used for this purpose. You will need these data later.

3. Into a 10-mL Erlenmeyer flask (Figure 1.6) measure 4 mL of 1 M sodium carbonate solution. Into one 10 × 75-mm culture tube measure, using a graduated pipette (Figure 1.7), 1 mL of 1 M calcium chloride solution. Into another measure 1 mL of 3 M sulfuric acid solution. If pipettes are not supplied with the reagents, wash your pipette between transfers. Weigh all three of these together on the top-loading balance (Figure 1.8) by placing them in a 50-mL beaker.

4. Pour the calcium chloride solution into the flask, swirl the contents to mix them, note what happens, and then weigh all three containers again on the top-loading balance.

5. Remove the beaker from the balance. With the flask still in the 50-mL beaker, **carefully** pour the sulfuric acid solution into the flask. Do this as slowly as possible. Swirl the flask, note what happens, and after all evidence of further change has ceased, weigh all three containers together again in the 50-mL beaker.

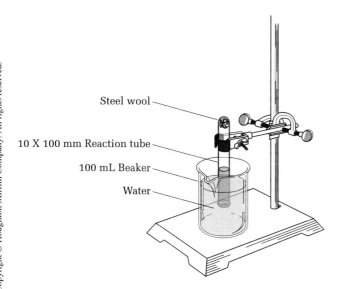

Figure 1.5 Apparatus for Step 2 of experiment.

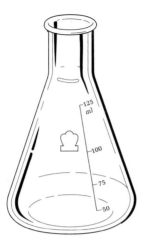

Figure 1.6 A 10-mL Erlenmeyer flask.

6. Estimate as closely as possible the decrease of the volume of air in the tube from Step 2. Record this volume in the appropriate blank in the Data and Observations section of the Report Sheet. If your tube is unmarked, record the length of the air column.

Weights and Masses

The mass of a substance is determined on a two-pan balance by comparison of the substance with a standard set of masses that come with the balance. The acceleration of gravity on the two sides of the balance is equal, so this comparison gives the mass of the substance (after correction for the buoyancy of air). But a spring-type bathroom scale does not make a comparison of this type. Its accuracy depends on the fact that the acceleration of gravity is *almost* the same all over the world. When you step on the scale you determine your weight, as determined by the force of gravity.

The modern one-pan electronic balance (Figure 1.8), which you will probably be using in this course, works by deflecting a small crystal, and hence it is, strictly speaking, determining weights, not masses. It has a calibration mass built into it; the balance must be calibrated as it is moved about on the surface of the earth because the force of gravity is not constant. Despite this you will often be directed to "Determine the mass of X" on the single-pan balance. The word *mass* is not a verb, so while you can *weigh* something you can't *mass* it.

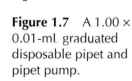

Figure 1.7 A 1.00 × 0.01-mL graduated disposable pipet and pipet pump.

How to Use the Top-Loading Balance

The electronic top-loading balance (Figure 1.8) is sensitive to a milligram (±0.001 g, ±1 mg) and is the most important and expensive piece of apparatus in the microscale laboratory. Treat it with respect.

Pressing the bar or the button on the front of the balance will turn it on and it will set itself to zero. The balance should be in a draft-free place or be provided with a removable draft shield. To weigh a container, simply place it carefully on the pan of the balance and then record the weight. This weight is referred to as the *tare* of the container. Add the material to be weighed and again record the weight. The maximum total weight (container plus contents) that can be placed on the balance is 160 g.

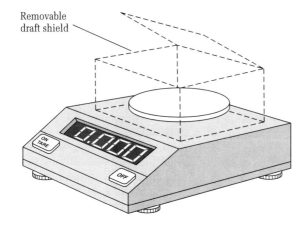

Figure 1.8 Top-loading balance, capacity 160 g, sensitivity ±0.001 g (±1 milligram).

Alternatively, if you do not need to know the tare of the container, it can be placed on the balance and then the balance zeroed by depressing the bar or button. Material to be weighed is added directly to the container until the desired weight is obtained.

Carefully brush off any material spilled on the pan or the top of the balance.

Cleaning Up

You first need to neutralize any acid remaining from the procedure. You will use 1.5 M aqueous sodium bicarbonate solution to carry out the neutralization. Two drops of 1.5 M bicarbonate solution will neutralize 1 drop of 3 M acid, either acetic or hydrochloric; 4 drops of the bicarbonate would be needed for each drop of the sulfuric acid, but most or all of it was consumed in Step 5.

Carefully remove the steel wool from the tube with forceps or tweezers (do not touch it with your hands). Place it in a small beaker or other container. Carefully add about 1 mL of sodium bicarbonate solution; note the reaction and describe it in the Data and Observations section. When the reaction in the beaker has stopped, use forceps or a boiling stick to remove and rinse the steel wool bits. Discard them in the waste basket.

Pour the contents of the 10-mL Erlenmeyer flask down the sink or into a waste container and wash it with lots of water. Rinse the tubes with water and turn them on their sides to dry. Wash your hands thoroughly with soap and water before leaving the laboratory.

Name _____ Section _____

Lab Instructor _____ Date _____

EXPERIMENT 1   Observing a Chemical Reaction

PRELABORATORY QUESTION

1. Calculate the average, the deviation, and the average deviation for the following volume readings. Express the average (mean) and include the average deviation.

Readings (mL)	Deviation, mL
2.54	_____
2.60	_____
2.50	_____
2.52	_____
2.49	_____

   Average _____

   Average Deviation _____

DATA AND OBSERVATIONS

1. Appearance of steel wool plus hydrochloric acid _____

   Appearance of steel wool plus acetic acid _____

2. Initial reading of volume of water in reaction tube _____

   Final reading of volume of water in reaction tube _____

   Total volume of reaction tube _____

   Percent decrease in volume in reaction tube _____

   Note: If you used an unmarked tube, calculate the percent decrease in the length of the trapped air column in the tube. If the tube were a true cylinder, this should give the same result as the percent volume decrease. (Why?) Discuss the validity of this assumption for the tube you used.

3. Mass of beaker, Erlenmeyer flask, culture tubes, and contents, weighed to the nearest milligram _____

   Appearance of solutions before mixing _____

   Mass of beaker, Erlenmeyer flask, culture tubes, and contents, weighed to the nearest milligram, after mixing calcium chloride, $CaCl_2$, with sodium carbonate, Na_2CO_3 _____

9

Appearance of solutions after mixing _____

Mass change, if any _____

4. Mass of beaker, Erlenmeyer flask, culture tubes, and contents, weighed to the nearest milligram, after mixing calcium chloride, $CaCl_2$, with sodium carbonate, Na_2CO_3, and then sulfuric acid, H_2SO_4 _____

   Appearance of solutions _____

   Mass change, if any _____

ANALYSIS

1. Obtain from your instructor a typical set of class results.
   a. Was the height of the water in the reaction tube the same for each trial? How consistent were the values for the percent of decrease in volume of air in the tube? Discuss the significance of any variations in results and offer an explanation for those variations.

   b. Were your observations on the calcium chloride, sodium carbonate, sulfuric acid reactions the same as those reported by others? Write the formulas for sodium carbonate and sodium bicarbonate.

   c. Do these formulas help you to explain the similarities of your observations in Step **6** and the **Cleaning Up** section? Discuss.

2. Report your results to your instructor and then calculate the average amount of gas left in the tube for the class. Also calculate the average deviation in this number.

CONCLUSIONS

1. Write your interpretation of your observations for each experiment. Explain how your observations make you reach the stated conclusion for each experiment. If you already know some chemistry, you might speculate on the chemical reactions involved, but remember that you have not *proven* that these reactions are occurring.

2. Compare your data with that of your classmates. Can you reach any further conclusions based on a large amount of data?

3. Suggest further experiments that might be done to confirm your conclusions and speculations.

POSTLABORATORY QUESTIONS

As you write conclusions you might think about some of the following questions:

1. Does bubbling indicate gas coming off or does it indicate boiling? How might you distinguish between the two?

2. Did the rise of water in the tube agree throughout the class (to ±0.25 mL)? _____

3. Did the steel wool change in appearance at the end of the experiment? _____

4. If the steel wool had been wet with 3 M hydrochloric acid, what might have happened to the water levels in the inverted reaction tube experiment?

5. If the steel wool in the reaction tube is perfectly dry, the results of this experiment are rather different. How does this relate to leaving your bicycle out in the rain?

6. Did your classmates all get the same results when calcium chloride solution was poured into sodium carbonate solution?

7. Did your classmates all make the same quantitative observations when sulfuric acid was added to the Erlenmeyer flask containing sodium carbonate?

8. One of the fundamental scientific principles involved in this experiment is the Law of Conservation of Mass.
   a. State the Law of Conservation of Mass.

   b. The famous French scientist, Antoine Lavoisier (who was later a victim of the French Revolution) is said to have carried out an experiment in which he weighed some paper, then set fire to it, then weighed the ashes left from the combustion. Would one expect to get the same mass both times in this experiment? Why (not)?

EXPERIMENT 2

Densities of Organic Liquids

Introduction

All substances—gases, liquids, and solids—exhibit the fundamental property of *density,* defined as mass per unit volume. The density of a substance changes with temperature. Water has its maximum density at 4°C, 1.00000 g/cm^3. At room temperature, 25°C, it is very slightly less, 0.9970 g/cm^3. The density of ice at 0°C is 0.9168 g/cm^3. The density of a substance is a constant at a given temperature and pressure. The density of a gas changes to an enormous extent with pressure, but liquids and solids have relatively constant densities as a function of pressure. The density of hydrogen, 0.0899 g/L, is less than that of dry air, 0.9982 g/L, at room temperature, as is evident from the behavior of hydrogen-filled balloons. One of the least dense solid elements at room temperature is lithium (0.534 g/cm^3), and osmium is the most dense (22.5 g/cm^3).

To determine the density of a liquid, we need to measure the mass and volume of the substance. The mass is measured on a balance and the volume in a calibrated volumetric device such as a pipette, a burette, or a volumetric flask. We will use a 1.00-mL graduated pipette. The liquids to be used in this experiment are common organic solvents, some of which you may recognize by their odors. Table 2.1 on page 15 lists several possibilities.

Precision, Accuracy, and Average Deviation

Precision refers to the closeness of repeated measurements to a common value, whereas *accuracy* is the closeness of a measurement to the true value (Figure 2.1).

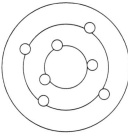
Accurate, but not precise

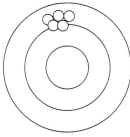
Precise, but not accurate

Figure 2.1 The difference between precision and accuracy applied to a shooting target.

To estimate the precision of a group of measurements, calculate the *average deviation* of the measurements. This is done by determining the average, or *mean*, of the measurements by adding up the individual measurements and dividing by the number of measurements. Then determine the absolute deviation (disregard the sign) of the average from each individual measurement. The average of these absolute deviations is the *average deviation*.

Experimenter	Density (g/mL)
1	0.98
2	0.96
3	0.99
4	0.95
5	0.97
Total	4.85

$$\text{Average} = \frac{4.85}{5} = 0.97 \text{ g/mL}$$

Experimenter	Density (g/mL)	Absolute Deviation
1	0.98	0.01
2	0.96	0.01
3	0.99	0.02
4	0.95	0.02
5	0.97	0.00
		0.06

$$\text{Average deviation} = \frac{0.06}{5} = 0.01$$

So the density of the liquid is 0.97 ± 0.01 g/mL.

For each of the Steps 3 to 6, draw liquid into the pipette with the syringe near the top graduation, but do not try to bring it to 1.00 exactly. Record the volume to ±0.001 mL. Then expel about 0.9 mL of the liquid into the vial. Again record the volume to ±0.001 mL. Do not try to adjust the liquid level to any particular line on the pipette. In recording the volume, read the bottom of the meniscus (the curved surface of the liquid). You can easily read it to two figures; you should try to estimate to the third decimal place. To avoid heating up the pipette with your hands, grasp the apparatus by the syringe.

Procedure Summary

You will determine the densities of four liquids: three whose identities are known, plus one unknown. The densities of the knowns will be compared with the accepted values shown in Table 2.1. As part of your report, you will conduct an extensive analysis of your results and those of your classmates. To do this, you will calculate the relative error of your own determinations and the average deviation shown by yourself and several others.

Table 2.1 Densities of Some Organic Liquids (g/mL)

Pentane, C_5H_{12}	0.626
Hexane, C_6H_{14}	0.659
tert-Butyl methyl ether, $CH_3OC(CH_3)_3$	0.741
Toluene, $C_6H_5CH_3$	0.867
Ethyl alcohol, CH_3CH_2OH	0.785
tert-Butyl alcohol, $(CH_3)_3COH$	0.786
Ethyl acetate, $CH_3COOCH_2CH_3$	0.902
Acetone, CH_3COCH_3	0.791
Diiodomethane, CH_2I_2	3.325
Carbon tetrachloride, CCl_4	1.594
Chloroform, $CHCl_3$	1.492
Dichloromethane, CH_2Cl_2	1.325

Prelaboratory Assignment

Read the Introduction and Procedure sections and answer the Prelaboratory Questions on the Report Sheet before you come to the laboratory.

Materials

Apparatus

Milligram balance
1-mL syringe, graduated
1.0 × 0.01-mL pipette, TC or TD
Sample vials with caps
2-cm. length of small-diameter gum rubber tubing
Reaction tube or small culture tube
Safety goggles
Laboratory apron

Reagents

Ethyl acetate
Hexane
Acetone
Unknown liquid

Safety Information

1. **Safety goggles must be worn at all times in the laboratory.** Contact lenses should not be worn when organic vapors are present.
2. **No open flames in the laboratory.** Organic liquids are flammable.
3. **Do not breathe the vapors.** Work only in a well-ventilated space.

Procedure

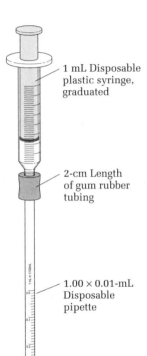

Figure 2.2 Graduated pipette delivery system.

- 1 mL Disposable plastic syringe, graduated
- 2-cm Length of gum rubber tubing
- 1.00 × 0.01-mL Disposable pipette

Record all data in the data table on the Report Sheet.

1. First examine the pipette carefully. Two types are commonly used. One is labeled "TC" at the top, which means "to contain," and the other is labeled "TD," which means "to deliver." The TD pipette delivers 1.00 mL from the top graduation to the bottom of the pipette, whereas the TC pipette delivers 1.00 mL from the top graduation to a bottom graduation, marked "0." The small volume below the zero mark is unknown. You will probably be using a TD pipette.

2. Attach a 1-mL plastic syringe to the pipette using a 2-cm length of gum rubber tubing as seen in Figure 2.2. Determine the masses of four dry, labeled, and capped vials (Figure 2.3) to the nearest milligram (0.001 g) on a top-loading balance.

3. Using the syringe as a pump, draw liquid into the pipette. You should try to get the liquid near the top graduation, but it is not necessary to hit 1.00 exactly. Reading the bottom of the meniscus (the curved surface of the liquid), record the volume of liquid in the pipette to ±1 μL (0.001 mL). You can easily read it to two figures; you should try to estimate the third figure, for example, 0.894 mL, but bear in mind that the last figure is not precise. To avoid heating up the pipette with your hands, grasp the apparatus by the syringe.

    Now expel about 0.9 mL of the liquid into the vial. Again record the volume to ±1 μL. As before, you need not hit any particular line on the pipette, but be sure that the liquid level is above the zero mark if you are using a TC pipette.

    Pipette into the first vial about 0.9 mL of hexane, C_6H_{14}, and cap the vial. Remember to record the beginning and ending volumes to ±0.001 mL (±1 μL). Some liquid should be left in the pipette. Add 1 or 2 very small drops of that liquid to about 10 drops of water in a reaction tube or 10 × 75-mm culture tube. What do you observe? Does the liquid dissolve in the water? If not, does it rise or sink in the water? Empty the pipette completely into a waste flask at your desk and allow the pipette to drain dry into a piece of filter paper.

4. Using a second vial, repeat Step 3, but substitute about 0.9 mL of ethyl acetate, $CH_3COOCH_2CH_3$, for the hexane. Make and record the observations described in Step 3.

5. With the third vial, repeat Step 3, but substitute about 0.9 mL of acetone, CH_3COCH_3, for the hexane. Make and record the observations described in Step 3.

6. Into the fourth vial, pipette about 0.9 mL of an unknown organic liquid, supplied by your instructor. Cap the vial. Again add 1 very small drop of the unknown to about 10 drops of water in a reaction tube or culture tube. Does the liquid dissolve in the water? If not, does it rise or sink in the water? Rinse the pipette with a bit of ethyl alcohol, then allow it to drain dry onto a piece of filter paper.

7. Determine the masses of the four vials plus the liquids in them to ±0.001 g (±1 mg).

Figure 2.3 Capped vial.

Cleaning Up

Empty your three vials into the container labeled "Organic Waste" and empty the vial containing the unknown and the water used to test its density and solubility into the container labeled "Unknown Waste."

8. Calculate the relative error for your first three measurements (see the following section).

Relative Error Calculation

Determine the masses of the four vials plus the liquids in them to ±0.001 g (±1 mg).

Calculate the relative error (percentage error) for your first three densities. This is done by determining the difference between your calculated value for the density and that reported in Table 2.1. Relative error can be expressed in various ways. For example, if the density you determined for a liquid was 0.91 g/mL and the value reported in Table 2.1 is 0.93 g/mL then the difference is 0.02 g/mL. When this is divided by the reported value, a fraction without units is obtained (0.02 g/mL ÷ 0.93 g/mL = 0.022). Multiplication of this number by 100 gives 2.2 parts per hundred, usually expressed as 2.2%. Multiplication by 1,000 gives 22 parts per thousand, or 22 ppt.

From the information in Table 2.1, identify your unknown.

Name _____ Section _____

Lab Instructor _____ Date _____

EXPERIMENT 2 Densities of Organic Liquids

PRELABORATORY QUESTIONS

1. To how many significant figures is the density of water given (in the Introduction) at 4°C? At 25°C?

2. On one pan of a big two-pan balance is placed a 1-L flask that has in the neck a very tightly fitting rubber stopper. On the other pan brass weights are added until perfect balance is achieved. The whole apparatus is placed in a vacuum chamber and the air removed from the chamber. Now the balance is tipped in favor of the flask. It seems to weigh more. Why?

3. Determine the relative error for a density experiment in which the accepted value is 0.750 g/mL, and the experimentally obtained value is 0.735 g/mL.

DATA AND OBSERVATIONS

1. Hexane

   Mass of empty vial and cap   _____

   Mass of vial and cap plus liquid _____

   Mass of liquid   _____

   Initial reading on pipette   _____

   Final reading on pipette   _____

19

Volume of liquid _____

Does it appear to dissolve in water? _____

If it does not dissolve, does it float or sink? _____

2. Ethyl acetate

   Mass of empty vial and cap _____

   Mass of vial and cap plus liquid _____

   Mass of liquid _____

   Initial reading on pipette _____

   Final reading on pipette _____

   Volume of liquid _____

   Does it appear to dissolve in water? _____

   If it does not dissolve, does it float or sink? _____

3. Acetone

   Mass of empty vial and cap _____

   Mass of vial and cap plus liquid _____

   Mass of liquid _____

   Initial reading on pipette _____

   Final reading on pipette _____

   Volume of liquid _____

   Does it appear to dissolve in water? _____

   If it does not dissolve, does it float or sink? _____

4. Unknown

   Mass of empty vial and cap _____

   Mass of vial and cap plus liquid _____

   Mass of liquid _____

   Initial reading on pipette _____

   Final reading on pipette _____

Volume of liquid   _____

Does it appear to dissolve in water?   _____

If it does not dissolve, does it float or sink?   _____

ANALYSIS AND RESULTS

Calculate the densities of the three liquids using the relationship

$$\text{Density (g/mL)} = \frac{\text{mass (g)}}{\text{volume (mL)}}$$

DATA TABLE

Your results for the densities of:

Hexane   _____   Acetone   _____

Ethyl acetate   _____   Unknown   _____

Record the densities in the Data Table. Pay particular attention to the number of *significant figures* in your calculations. Although the balance may give a reading such as 5.678 g (four significant figures), you can read the pipette to only three significant figures (e.g., 0.894 mL). When you calculate the density of the liquid you may end up with a ratio such as 0.806 g/0.894 mL, which in your electronic calculator will come out 0.90156599553 g/mL (11 significant figures), but this *must* be rounded off to just three significant figures (0.902 g/mL), because the least precise measurements in the experiment were made to just three significant figures. Enter the results of your calculations below.

1. Calculated density of hexane   _____

   Relative error   _____

2. Calculated density of ethyl acetate   _____

   Relative error   _____

3. Calculated density of acetone   _____

   Relative error   _____

4. Calculated density of unknown   _____

   Probable identity of the unknown   _____

   Relative error in density determination,
   based on the above identification   _____

POSTLABORATORY QUESTIONS

1. a. Consult with four other members of the class to get their values for the density of ethyl acetate. List the values, including yours, then determine the average density (called the *mean*) for the liquid and the average deviation. Show your calculations below.

	Densities	Deviations
1	_____	_____
2	_____	_____
3	_____	_____
4	_____	_____
5	_____	_____
Totals	_____	_____

   Average Density _____

   Average Deviation _____

   b. How closely does the average deviation for five different experimenters correspond with your own percentage of relative error for ethyl acetate?

   c. Which of the two figures (your relative error or the average percent deviation) would be of more value to someone who was interested in knowing the physical properties of ethyl acetate? Explain.

2. Would the calculated densities for these liquids be larger or smaller if some of the liquid evaporated before the vial could be capped? Explain.

3. Why attach a pipette to the syringe? The syringe also holds 1 mL and is calibrated. Why not just use it?

EXPERIMENT 3

Separation of a Mixture

Introduction

Polyethylene

Urea

Naphthalene

Chemistry is the study of the chemical and physical properties of substances and their transformations. To carry out these studies, the chemist must deal with pure substances, so a large part of a chemist's work is associated with the separation of mixtures and the purification of substances.

Have you ever wondered how gasoline, petroleum jelly, and asphalt are obtained from crude oil or how, when a soft drink bottle is recycled, the aluminum cap, the paper label, the dark plastic bottom and the clear or green body of the bottle are separated? Separations are carried out by taking advantage of the physical and chemical properties of the substances in the mixture.

In this experiment you will be given a mixture of high-density polyethylene, $(CH_2CH_2)_n$ (the translucent plastic used to make gallon milk containers); iron, Fe; sand, SiO_2; urea, NH_2CONH_2; and naphthalene, $C_{10}H_8$ (mothballs). You will separate this mixture using the various physical properties of the materials. You can probably guess how you will remove the iron, but clean separation of the other substances may not be as obvious.

You will use the physical properties of magnetism, vapor pressure, solubility, and density to separate these materials. Examine the mixture closely and record your observations. With enough patience you could probably pick the components apart under a microscope, but this would not be practical if you had a ton of material to separate.

Note the color of the mixture. You may not know the color of urea (it is white), but sand, iron, and mothballs should be familiar to you. The polyethylene could be almost any color; its natural color is a translucent white.

Note the odor of the mixture. What odors do sand and iron and polyethylene have? Again, you might not know about urea (it has no odor). What do mothballs smell like? Do mothballs stay around forever? What happens to them?

Procedure Summary

In this experiment the iron will be removed from the mixture with a magnet, the naphthalene (mothballs) by sublimation, the urea by solution in water, and the polyethylene by flotation on water and filtration. The sand will be separated from the water by filtration.

This experiment is an exercise in careful laboratory technique. Try to separate the components without loss.

23

Experiment 3 Separation of a Mixture

Prelaboratory Assignment

Read the Introduction and Procedure sections carefully and answer the Prelaboratory Questions on the Report Sheet.

Propose some ways for separating the components. Look up the properties of the components of the mixture in a reference book such as *The CRC Handbook of Chemistry and Physics*. In particular, note the solubilities of the components in various solvents.

Materials

Apparatus

Sublimation apparatus
   Filter flask
   Pipette bulb
   Centrifuge tube with Pluro stopper
Hirsch funnel
Stand and clamp to support filter flask
Water aspirator or other source of vacuum (for example, a 50-mL plastic syringe)
Ice
Magnet
Glazed paper, 15 × 15 cm (a magazine cover)
Steam bath or hot plate and beaker of hot water or electrically heated sand bath
Wooden applicator stick
100-mL beaker
10-mL Erlenmeyer flask
Pasteur pipette or Beral pipette
Filter paper
4 × 4-cm polyethylene bags for separated products

Reagents

A mixture of iron, naphthalene, polyethylene, urea, and sand

Safety Information

1. **Safety goggles must be worn at all times in the laboratory.**
2. **The sand bath is hot. Do not touch the sand.**
3. **Contact with steam can cause severe burns.**

Procedure

1. Obtain bags for your products, then place onto a piece of weighing paper approximately 2.5 g (±0.001 g) of the mixture. Place the entire mixture on a 15 × 15-cm square of heavily glazed or waxed paper. Using a magnet under the paper, move the iron powder into a pile. It is best not to allow the iron particles to touch the magnet directly because they are very hard to remove and will carry other material with them. Scrape the pile of iron powder onto a tared (previously weighed) piece of paper and determine the mass of the iron. Examine it carefully. Is it pure? Place it in a 4 × 4-cm polyethylene bag to be handed in with your report.

2. Weigh the clean, dry, empty filter flask and record its mass. You will need this information later. Place the mixture remaining after the removal of iron in the filter flask. Close the sidearm of the flask with a rubber bulb. Place the 15-mL glass centrifuge tube, which has been fitted with an adapter, in the neck of the flask. Add ice or very cold water to the 15-mL centrifuge tube, cap the tube to prevent spills, and heat the filter flask in a steam bath or a 100-mL beaker of hot water (Figure 3.1). Note very carefully what happens inside the flask. You are witnessing the process of *sublimation*.

    After no further change is noted, remove the flask from the steam bath or beaker of boiling water, cool the flask to room temperature, empty the cold water from the centrifuge tube, and replace it with water at room temperature.

3. Gently wipe all water drops from the top of the flask, adapter, and centrifuge tube and carefully remove the tube from the filter flask. Scrape the crystals onto a piece of tared paper and determine their mass. Place the crystals in a polyethylene bag to be handed in with the report.

4. Transfer the material left in the filter flask to a 10-mL Erlenmeyer flask or a small test tube. Add 1 mL of water to the mixture and stir the contents with a wooden applicator stick.

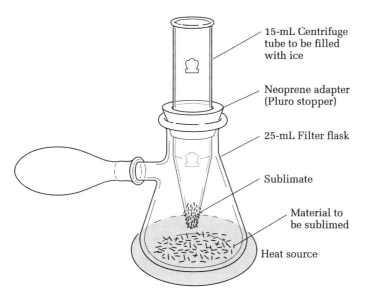

Figure 3.1 Sublimation apparatus.

5. Then place the Hirsch funnel with the fritted filter disk in the filter flask and add a 12-mm piece of filter paper to keep the frit clean. Remove the pipette bulb from the sidearm and attach the rubber hose to the aspirator. Clamp the flask so it does not tip over (Figure 3.2) and connect it through a trap to the water aspirator. Turn on the water in the aspirator in order to generate a vacuum in the filter flask. Transfer the aqueous solution of urea to the Hirsch funnel with a Pasteur pipette or a plastic Beral pipette. Rinse the Erlenmeyer flask and its contents with a few drops of water and filter that through the Hirsch funnel also. Try to leave the polyethylene and sand behind.

6. Evaporation of the water from the urea solution in the filter flask will leave crystalline urea behind. To do this, warm the filter flask on a steam bath or hot water bath while a vacuum is being applied to the flask. Using your thumb in the opening of the Hirsch funnel to control the vacuum (Figure 3.3), evaporate the water from the solution to leave urea in the flask.

    Wipe away any droplets of water from the neck of the flask. Determine the mass of the urea by weighing the flask plus urea and subtracting the weight of the flask, determined previously. Scrape as much of the urea as possible into a polyethylene bag to be turned in with the report.

    Another way to get rid of the water is by evaporation. Pour the water solution on a tared watch glass and set it out of the way in your locker for a week. The water will evaporate leaving the urea behind. Weigh the watch glass plus urea and subtract the tare of the watch glass to determine the weight of the urea.

7. Add water to the mixture of polyethylene and sand left in the 10-mL Erlenmeyer flask. Decant (pour off) the top layer of polyethylene and water into the Hirsch funnel, leaving behind the much denser sand in the Erlenmeyer flask.

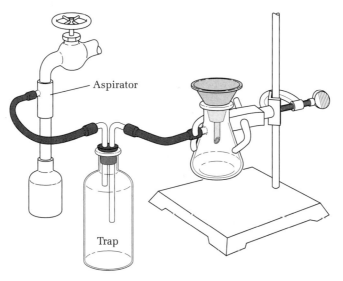

Figure 3.2 Filtration assembly.

Figure 3.3 Evaporation of a liquid under vacuum.

Apply a vacuum to the filter flask and get the polyethylene as dry as possible. Scrape it out on to a piece of filter paper, blot it dry, determine its mass, then bag it.

8. Scrape the sand from the flask on to the Hirsch funnel. Apply a vacuum to dry it as much as possible and then scrape the sand onto a piece of filter paper, blot it dry, and determine its mass. Bag it.

9. Find the total weight of the separate components of the mixture. What percent of the original mixture did you recover? If the total mass is more or less than what you started with, account for the discrepancy.

Cleaning Up

There should be nothing to discard from this experiment. Since nothing should be lost during this experiment, everything can be recycled.

Name _____ Section _____

Lab Instructor _____ Date _____

EXPERIMENT 3   Separation of a Mixture

PRELABORATORY QUESTIONS

1. Look up the properties of the components of the mixture in a reference book such as the *CRC Handbook of Chemistry and Physics*. In particular note the solubilities of the components in various solvents. Which, if any, are soluble in water? Which are not? How could you take advantage of these solubility differences?

2. Look back at the discussion of moth balls in the Introduction.

    a. What does happen to moth balls over a period of time?

    b. How can you tell they are present without actually seeing them?

    c. Offer an explanation for both of these phenomena. (Hint: What must happen to them?)

3. Consider the five solids used in this experiment.

    a. Are any of them elements? If so, which one(s)?

    b. Which, if any, might be considered mixtures in their own right?

Experiment 3   Report Sheet

DATA AND OBSERVATIONS

Mass of original mixture   _____

Mass of iron   _____

Mass of clean, dry filter flask   _____

Mass of naphthalene   _____

Mass of filter flask and urea   _____

Mass of filter flask (from above)   _____

Mass of urea   _____

Mass of polyethylene   _____

Mass of sand   _____

Total mass of recovered material   _____

Percent recovery   _____

CALCULATIONS

$$\text{Percent recovery} = \frac{\text{mass recovered}}{\text{initial mass of mixture}} \times 100\%$$

CONCLUSIONS

1. Can any mixture of substances be separated cleanly by the process outlined here? What is the principal difficulty encountered with this process in this experiment?

POSTLABORATORY QUESTIONS

1. Is the order in which this separation is carried out important? For instance, what problem might be encountered if the urea were removed first?

2. Suggest two ways to separate naphthalene from urea.

3. From Experiment 2, *Densities of Organic Liquids*, can you predict whether naphthalene and polyethylene, both insoluble in water, will float or sink in water? Obviously, they are not liquids, but they are both hydrocarbons, like hexane.

4. In gold rush days, miners panning for gold separated bits of gold from stream beds from the sand and gravel by swirling water through the mixture. Sand was washed away, leaving the gold behind. To what part of the Procedure is this analogous? What physical property of gold and sand is responsible for the separation? Similarly, in the days when people raised and harvested their own wheat, the wheat grains were separated from the chaff (hulls) by tossing everything up in the air and catching it as it fell back. The wheat grains were caught in the basket, while the chaff was blown away.

EXPERIMENT

4 Crystallization

Introduction

Crystallization is a method for purifying substances that is carried out on an enormous scale industrially. Sugar and salt are two examples of substances purified in this way. The object of this experiment is to observe the process of crystallization and to purify a chemical compound by this process.

When a *solute,* such as ordinary table salt, is dissolved in water, the *solvent,* a clear solution, is obtained. When no more salt will dissolve in a given volume of water the solution is said to be *saturated*. In general, a hot solvent will dissolve more of a solute than a cold solvent (Figure 4.1).

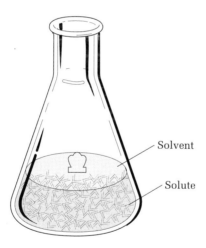

Figure 4.1 Solute and solvent in Erlenmeyer flask.

Crystallization is the process whereby crystals form from a saturated solution of a solute. In one method of crystallization, a solution of the solute is evaporated to deposit the solute as crystals, a process used for millennia to obtain salt from seawater.

In another process a hot saturated solution is cooled. As it cools crystals come out of solution. This is an important means of purification because if the solution also contains an impurity that does not saturate the solution, then the impurity stays in solution as the desired substance crystallizes. White sugar is obtained from a solution of brown sugar by this method.

In the first experiment a mixture of two substances will be separated by crystallization from water. One is much less soluble in the water than the other and so it will crystallize from a hot solution. The other substance will stay in solution.

34   Experiment 4   Crystallization

In the second experiment an aspirin tablet will be separated from its binder and then crystallized from a mixture of solvents.

Occasionally, a substance will not crystallize from a hot saturated solution as the solution is cooled. The solution is said to be *supersaturated*. Addition of a seed crystal will initiate crystallization.

Procedure Summary

Two compounds are purified by crystallization. Impure phthalic acid is recrystallized from hot water, and the aspirin in an aspirin tablet is dissolved in a very good solvent, the solution filtered to remove the binder used in the tablet, and then a poor solvent added to cause the aspirin to crystallize in pure form.

Prelaboratory Assignment

Read the Introduction and Procedure sections carefully and answer the Prelaboratory Question on the Report Sheet.

Materials

Apparatus

Electrically heated sand bath
10 × 100-mm reaction tube
Wood boiling stick
Hirsch funnel
12-mm dia. filter paper
25-mL filter flask
Pasteur pipette and bulb
Clamp and ringstand to support filter flask
Cotton or paper towel (for insulation in Step 3)

Reagents

Impure phthalic acid
Aspirin tablets
tert-Butyl methyl ether
Hexane
Ice, for ice water bath

> **Safety Information**
> 1. **Safety glasses must be worn at all times in the laboratory.**
> 2. **The sand bath is hot.** Avoid touching it.
> 3. When boiling the solution in the reaction tube, point the opening of the tube away from yourself and your lab mates in case the liquid should "bump" (boil violently out of the tube). A boiling stone or as illustrated in Fig. 4.2, a boiling stick, will help prevent bumping.

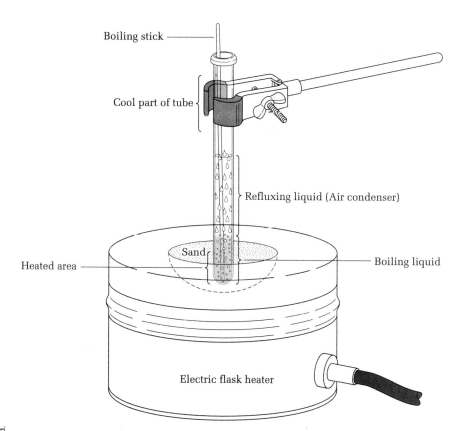

Figure 4.2 Heating a reaction tube on an electrically heated sand bath.

Procedure

Phthalic acid

Recrystallization of Impure Phthalic Acid

1. Turn on your electrically heated sand bath (Figure 4.2) to a setting of 3 on the controller. Place about 100 mg of the red powder (impure phthalic acid) and a boiling stick (which promotes smooth boiling) in a reaction tube or 75-mm test tube and add to it 1.0 mL of water. Heat the mixture carefully to boiling on the hot sand bath. Note carefully what happens to the red powder.

2. Cool the solution rapidly in a beaker of ice water with shaking and stirring of the solution. Note the appearance of the mixture once it has cooled.

3. Reheat the red mixture until a clear solution is obtained. This time remove the boiling stick and wrap the reaction tube or test tube in a wad of cotton, a paper towel, or some other insulating material. Allow the solution to cool undisturbed for several minutes. Then, without shaking the solution, place it gently in a beaker of cool water to continue cooling. Note the appearance of the contents of the tube this time.

4. Using a Pasteur pipette, remove the red filtrate (called the *mother liquor*) from the crystals. To do this, gently push air out of the pipette as it is being pushed down through the crystals. Seat the tip of the pipette squarely on the bottom of the reaction tube (see Figure 4.3), withdraw the solution into the pipette, and then squirt it into another reaction tube. Add about 1 mL of ice water to the crystals and again withdraw the filtrate. Repeat this process once more.

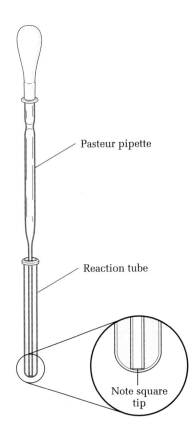

Figure 4.3 Filtration of a solution using a Pasteur pipette.

5. Using the metal spatula scrape some of the crystals out of the tube onto a piece of filter paper. Observe and record their appearance.

Cleaning Up

The crystals, the contents of the reaction tube, and the red filtrate should be placed in the appropriate waste container.

Crystallization of Aspirin (Acetylsalicylic Acid)

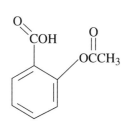

Acetylsalicylic acid

Commercial aspirin is acetylsalicylic acid, but the pure substance, when fashioned into pills, has a tendency to fall apart or to be so hard that they do not dissolve easily. For this reason these pills, and many others as well, contain a *binder*, which a close reading of the label will disclose. This can be an organic substance such as microcrystalline cellulose, or an inorganic substance such as silica. These binders are often insoluble in solvents, so when an aspirin tablet is dissolved in water or some other solvent the solution is not clear.

In the present experiment an aspirin tablet will be dissolved in an organic solvent in which it is very soluble [*tert*-butyl methyl ether, $(CH_3)_3COCH_3$]. The resulting cloudy solution will be filtered to remove the binder. To the filtrate, the liquid that has been filtered, is added a different solvent [hexane, $CH_3(CH_2)_4CH_3$] in which aspirin is not very soluble. Pure aspirin will crystallize from this mixture of solvents.

6. Weigh an aspirin tablet and then in a reaction tube dissolve it in the minimum quantity of hot *tert*-butyl methyl ether on a steam bath or a warm sand bath. Add a boiling stick to promote even boiling. Note the appearance of the mixture. Add more *t*-butyl methyl ether equal to half that used initially and then filter the solution through the Hirsch funnel that has been equipped with a 12-mm piece of filter paper (Figure 4.4).

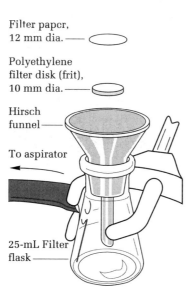

Figure 4.4 Filtration using the Hirsch funnel.

7. Wash out the reaction tube with a few drops of solvent and use these drops to rinse the Hirsch funnel. Using a Pasteur pipette, transfer the filtrate to a reaction tube and warm the tube on the sand or steam bath. Then add an equal volume of hexane and allow the tube to cool to room temperature. If the aspirin does not crystallize, seed the supersaturated solution with a minute speck of aspirin powder. Cool the reaction tube in ice, note the appearance of the crystals, then stir the contents of the tube and collect the recrystallized aspirin on the Hirsch funnel, using a new, clean piece of filter paper.

8. Once the aspirin is dry, determine its mass and calculate the percent recovery based on the weight of the original aspirin tablet. Place the recovered aspirin in a 4 × 4-cm polyethylene bag and attach it to your laboratory report.

Cleaning Up

Pour the final filtrate, a mixture of *t*-butyl methyl ether and hexane, into the organic waste container. Wash out the reaction tube, Hirsch funnel, and filter flask with water and allow them to drain dry.

Name _____   Section _____

Lab Instructor _____   Date _____

EXPERIMENT 4 Crystallization

PRELABORATORY QUESTION

1. From your experience and a knowledge of the process of crystallization, cite some other examples of compounds that might be purified by crystallization.

DATA AND OBSERVATIONS

Recrystallization of Impure Phthalic Acid

Appearance of powder at beginning of experiment _____

Appearance of initial solution of powder _____

Effect of heating suspension of powder _____

Appearance of mixture after being cooled rapidly _____

Appearance of mixture after being cooled slowly _____

Which method probably gives purer crystals? Explain your choice. _____

_____

Color of recrystallized phthalic acid _____

Color of filtrate _____

Crystallization of Aspirin (Acetylsalicylic Acid)

Mass of aspirin tablet _____

Appearance of the initial solution of aspirin in the solvent _____

Appearance of filtrate _____

Appearance of binder, on the filter paper _____

40 Experiment 4 Report Sheet

Is supersaturation detected? _____

If so, what evidence do you have? _____

Appearance of crystals of aspirin _____

Mass of dry recovered aspirin _____

Percent recovery of recrystallized aspirin _____

$$\frac{\text{Mass of recrystallized aspirin}}{\text{Mass of aspirin tablet}} \times 100 = \% \text{ recovery}$$

POSTLABORATORY QUESTIONS

1. Explain how a solution can be both saturated and dilute.

2. In Part I of the procedure, what assumption is being made about any impurities that may be in the original phthalic acid sample?

3. From your experience and a knowledge of the process of crystallization, cite some other examples of compounds that may have been purified by crystallization.

EXPERIMENT 5

Sublimation

Introduction

Sublimation is the process whereby a substance passes directly from the solid phase to the vapor phase, whereas the more familiar process of evaporation is the process whereby a substance passes from the liquid phase to the vapor phase.

Dry ice (solid carbon dioxide) sublimes as it disappears, as do bathroom deodorizers. In colder parts of the United States, you may notice that snow will disappear over a period of time, even though the temperature never goes above freezing. The snow has sublimed into water vapor without ever melting. It is quite possible to hang wet laundry out with the temperature below freezing. The clothes will freeze solid and then, over a period of time, will become completely dry as the water in them sublimes.

Procedure Summary

In the present experiment an impure solid will be sublimed in order to purify it. The impurity will not sublime. The purified material can be identified by its *melting point* (MP). The temperature of melting is a characteristic property of most organic solids. In general, if you can detect the odor of a pure solid chemical, then it can be sublimed because your nose is detecting molecules of the solid that have sublimed into the vapor phase.

Prelaboratory Assignment

Read the Introduction and Procedure sections carefully, noting especially the appearance of the distillation apparatus and the functions of the various parts. Answer the Prelaboratory Questions on the Report Sheet.

Materials

Apparatus

25-mL filter flask
Pluro stopper
15-mL glass centrifuge tube
Rubber pipette bulb
Electrically heated sand bath
Hair dryer or heat gun (optional)
Ice

Reagents

Naphthalene, dichlorobenzene, or benzoic acid
Unknowns 1, 2, or 3

Safety Information

1. **Safety goggles must be worn at all times in the laboratory.**
2. **The sand bath is hot.** Avoid touching it.
3. **Do not touch unknown chemicals.** Avoid contact with the unknowns.

Procedure

1. Measure 50 mg of one of the numbered unknown substances into the filter flask. Note its number and appearance. Fit the flask with a rubber bulb on the sidearm and then add a centrifuge tube that has been fitted with a Pluro stopper (rubber adapter). Once the apparatus has been assembled, put ice into the centrifuge tube. Gently warm the filter flask on the hot sand bath (Figure 5.1). Note carefully what happens. Continue warming the filter flask and roll it in the sand so that the material on the upper walls of the flask is transferred to the centrifuge tube. One of the best ways to heat the filter flask and make all of the unknown sublime onto the centrifuge tube is to heat the flask with a heat gun or a gun-type hair dryer. Do not do this with the apparatus in the sand bath; you will blow away the sand.

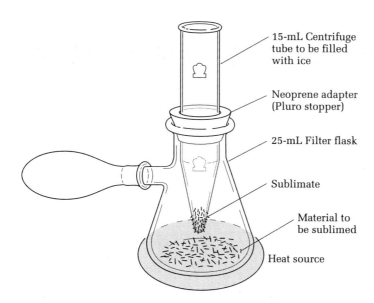

Figure 5.1 Sublimation apparatus

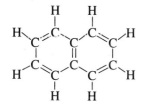

Naphthalene

Benzoic Acid

p-Dichlorobenzene

2. When the sublimation is complete allow the filter flask to cool to near room temperature, empty the ice water from the centrifuge tube, and then fill the centrifuge tube with water at room temperature. Only then can you remove the centrifuge tube without having moisture from the air collect on the cold tube.

   Note the appearance of the sublimed material and of the residue in the filter flask. Should any material fall from the centrifuge tube back into the flask when you are separating the two simply turn the filter flask upside down over a clean sheet of paper. Usually the sublimed material will fall from the flask and the impurity will stick to the bottom of the filter flask.

3. Determine the mass of the sublimed material and calculate a percent recovery based on the weight of the crude material.

You may be asked to identify your unknown by its melting point. Crude melting points can be determined by figuring out if the substance melts on the surface of a steam bath. If it does not, it is benzoic acid. The other two substances can be distinguished by touching them to a flask or tube of water heated to about 70°C. Only the *p*-dichlorobenzene will melt.

Table 5.1 Possible Sublimation Unknowns

Substance	MP (°C)
Naphthalene	81–83
p-Dichlorobenzene	54–56
Benzoic acid	122–123

Cleaning Up

Place the sublimed material in the organic waste container or, if it is *p*-dichlorobenzene, in the halogenated organic waste container.

Name _____ Section _____

Lab Instructor _____ Date _____

EXPERIMENT 5 Sublimation

PRELABORATORY QUESTIONS

1. Outline the order for assembling and taking apart the sublimation apparatus.

2. From your experience and a knowledge of the process of sublimation, cite some examples of substances that sublime.

3. How might freeze-dried foods be prepared?

DATA AND OBSERVATIONS

Unknown number _____

Weight of impure unknown _____

Appearance of impure unknown _____

Weight of sublimed material _____

Appearance of sublimed material _____

Percent recovery of sublimed material _____

Method of melting point determination _____

Approximate melting point of sublimed material _____

Experiment 5   Report Sheet

Is the melting point an acceptable way to identify the unknown? Explain. _____

Identity of unknown _____

What changes did you observe occurring in the filter flask during the sublimation? _____

_____

_____

Did anything enter or leave the flask during the sublimation? How do you know? _____

_____

_____

POSTLABORATORY QUESTIONS

1. Has sublimation aided in the purification of the unknown? Justify your answer.

2. If you did not get a high percent recovery of the sublimed product what might account for the missing material?

EXPERIMENT 6

Solubility and Solutions

Introduction

Virtually all chemical reactions occur in solutions. The earth's atmosphere is a solution of one gas in another; the oceans are solutions of salts in water. In living organisms complex reactions occur in solution. The purpose of this experiment is to explore the nature of solubility and of solutions.

A *solution* is a homogeneous mixture of substances that are uniformly dispersed in each other at the molecular level. Usually one component dominates. It is the *solvent*. The substance dissolved in the solvent is the *solute*.

In qualitative terms there are *dilute solutions*, in which the relative amount of solute dissolved in the solvent is small, and *concentrated solutions*, in which the amount dissolved is large. When no more solute will dissolve in the solvent at a given temperature the solution is *saturated*. In quantitative terms the concentration of the solute in the solvent is expressed in some convenient form such as 20 g/L, which is 20 grams of the solute dissolved in one liter of the solvent.

Not quite so common are *emulsions*, in which one substance is suspended in the other. Milk is an emulsion. Under a microscope we can see that milk consists of very small droplets of butterfat suspended in the aqueous phase. The fact that milk is not clear is an indication that it is not a true solution.

Chemists use solubility to characterize a large number of different substances. Solubility in a given solvent, like color, density, melting point, etc., is a fundamental property of a pure chemical substance. Solubility in various solvents is one of the first tests conducted on a new substance.

A very simple rule governs solubility: "like dissolves like." This means that solvents will dissolve substances that are similar to themselves—ionic solvents will dissolve ionic solutes, covalent solvents will dissolve covalent solutes.

In this experiment you will attempt to dissolve various solutes in both covalent and ionic solvents and test for dissociation of soluble solutes. In addition the effect of stirring, temperature, and subdivision of the solute will be studied.

Substances that form ions in water produce solutions that will conduct electricity and cause the light-emitting diode (LED) on the conductivity tester to light up. Some substances do not completely ionize and will only allow the LED to light dimly. Covalent substances do not ionize and so the solution will not conduct electricity.

Procedure Summary

Your objective in this experiment is to make careful observations of the solubility behavior of a number of different solutes in several solvents and to draw conclusions from those observations. Pay attention to temperature, particle size, stirring, the color of solutions, and the measurement of electrical conductivity.

Note carefully the distinction between what you observe and what you can then conclude. For example, does the observation that bubbles are rising from a solution mean that it is boiling or does it mean that a gas is being evolved?

Prelaboratory Assignment

Read the Introduction and Procedure sections, and answer the Prelaboratory Questions on the Report Sheet.

Materials

Apparatus

24-well test plate
Pasteur or Beral pipettes
Conductivity tester
Glass stirrers (melting point capillaries) or toothpicks
Micro vials, polypropylene or glass

Reagents

See the lists in the Procedure section.

Safety Information
1. **Safety goggles must be worn at all times in the laboratory.**
2. **Be sure there are no flames in the laboratory.** Organic liquids are flammable.

Experiment 6 Solubility and Solutions 49

Procedure

Part 1

1. In a 24-well polystyrene test plate (Figure 6.1) place just enough of the powder or crystals of the following substances so you can see them, except as noted below. The well plate has letters along one axis and numbers along the other, so you can record the location of samples, for example, C-3.
    a. Sodium bicarbonate
    b. Copper(II) nitrate trihydrate
    c. Nickel(II) nitrate hexahydrate
    d. Sulfur
    e. Sodium chloride
    f. Calcium chloride
    g. Ammonium chloride
    h. Silver chloride, a tiny speck (it is expensive)
    i. Zinc oxide
    j. Calcium oxide
    k. Silicon dioxide
    l. Sodium carbonate
    m. Potassium carbonate
    n. Calcium carbonate
    o. Calcium sulfate
    p. Copper sulfate pentahydrate
    q. Potassium permanganate, one small crystal
    r. Aluminum powder
    s. Zinc powder
    t. Tin powder
    u. Carbon, as charcoal
    v. Magnesium, one curl or small amount of powder
    w. Iron
    x. Sucrose

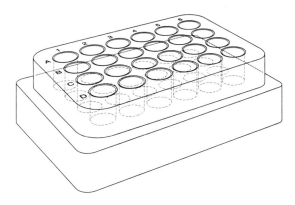

Figure 6.1 A 24-well polystyrene test plate.

2. From a plastic Beral pipette add a drop of water to each substance in the well plate, and stir with a very small glass rod (a melting point capillary). Note whether the substance dissolves. Wipe off the stirrer carefully each time it is removed from a well. Observe the well plate from above with an appropriate background, either white or black, behind each well. Add another drop of water to substances that did not dissolve and stir the substance again. Note any indication that a substance has dissolved. Note color changes in the solutions. Indicate whether or not each substance is water soluble by writing "yes" or "no" in the Data Table on page 54. Record any special observations for this or the next three steps in the "Observations" column in the Data Table.

3. Using a conductivity tester (Figure 6.2), touch each drop with the two leads and note whether the light-emitting diode (LED) lights up brightly, weakly, or not at all. Rinse with water and wipe the leads with a piece of absorbent paper after immersing in each drop of solution. Test to be sure it is clean by immersing the probe tips in distilled water. The LED should not light, because distilled water is a poor conductor. Indicate whether or not the substance conducts electricity by writing "yes" or "no" in the Data Table on page 54.

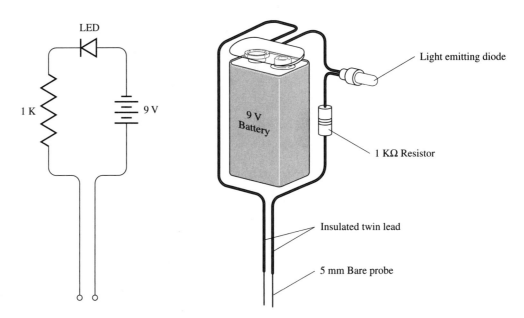

Figure 6.2 Conductivity tester.

4. Add one drop of 6 M hydrochloric acid (**caution!**) to each well and observe any indication of solution and/or reaction. Indicate whether or not the substance goes into solution when treated with acid by writing "yes" or "no" in the Data Table on page 54.

5. Add two drops of 6 M sodium hydroxide solution (**caution!**) to each well and again note any indication of solution and/or reaction. Because the first drop of sodium hydroxide neutralized the acid, the second drop should make each solution basic at this point. Indicate whether or not the substance goes into solution when treated with base by writing "yes" or "no" in the Data Table on page 54.

Experiment 6   Solubility and Solutions   51

Cleaning Up

Add a drop of 6 M hydrochloric acid to each well. At this point the solution in most of the wells should be neutral. Shake the 24-well test plate over the wide-mouth container provided. Most of the material in the plate will come out. Use a wash bottle to wash out material that does come out and then wash the well plate at the sink and invert it to dry.

Part 2

6. In a series of small, clear polyethylene vials or small glass shell vials (Figure 6.3) place 1 small drop or an equivalent amount (a very small amount) of the solid for each of the following substances. The clear plastic (polystyrene) well plate cannot be used because it will dissolve in acetone!
    a. Methanol
    b. 2-Propanol
    c. Ethylene glycol
    d. Cholesterol
    e. Naphthalene
    f. p-Dichlorobenzene
    g. Acetic acid
    h. Polystyrene (tiny piece of a white, disposable hot beverage container)
    i. Benzoic acid
    j. Sucrose
    k. Caffeine
    l. Citric acid
    m. Cyclohexane
    n. *tert*-Butyl methyl ether
    o. Ethyl acetate
    p. Iodine
    q. Vanillin
    r. Acetone

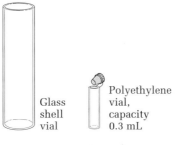

Figure 6.3 Container for solubility tests.

7. Add two drops of acetone to each substance, stir with a melting point capillary if necessary; observe and record in the first column of the Data Table on pages 55–56 whether or not the substance dissolves in (is miscible with) the acetone. Now add about 6 or 8 drops of water to each substance, in addition to the acetone that has been added already. Stir the mixture if necessary. Note the effect of stirring. Note any colors or color changes in the Data Table's observations column. Note whether the solution becomes cloudy, an indication that a substance dissolved in acetone is not soluble in water. Note whether

substances that did not dissolve in acetone will now dissolve in the water that has been added. Record whether or not the substance is water soluble by writing "yes" or "no" in the appropriate column in the Data Table. The last substance to be tested is acetone. It is on the list so that you can observe whether it is miscible with water. Test each solution with the conductivity meter.

Cleaning Up

Empty each of the tubes in the container provided for organic waste, except for the *p*-dichlorobenzene, which goes into the halogenated waste container. Many of these substances can also be disposed directly down the drain, flushing with water.

Name _____   Section _____

Lab Instructor _____   Date _____

EXPERIMENT 6 Solubility and Solutions

PRELABORATORY QUESTIONS

1. Look up the chemical formulas for the following common substances and correlate them with the substances in Part 1 of the experiment by giving each its chemical name.

    a. Limestone _____

    b. Lime _____

    c. Gypsum _____

    d. Table salt _____

    e. Sugar _____

    f. Baking soda _____

    g. Soda _____

    h. Sand _____

2. Look up the chemical formulas of the substances in Part 1 not identified in Prelaboratory Question 1.

Name	Formula	Name	Formula

3. Match the following with substances in the list in Part 2, #6 of this experiment:

    a. Causes the tart taste of lemons _____

    b. A stimulant _____

c. Has a bitter taste, but is a common "flavor" _____

d. Automobile antifreeze _____

e. Wood alcohol _____

f. A disinfectant, formerly widely used _____

g. An ingredient in fingernail polish remover _____

h. Antiknock additive in premium gasoline _____

i. Mothballs _____

j. Table sugar _____

k. Rubbing alcohol _____

l. Vinegar _____

m. Can clog arteries _____

n. A food preservative _____

DATA AND OBSERVATIONS

Part 1

	Location in Well Plate	Water sol?	Conductivity?	Acid sol?	Base sol?
a. Sodium bicarbonate	_____	_____	_____	_____	_____
b. Copper(II) nitrate trihydrate	_____	_____	_____	_____	_____
c. Nickel(II) nitrate hexahydrate	_____	_____	_____	_____	_____
d. Sulfur	_____	_____	_____	_____	_____
e. Sodium chloride	_____	_____	_____	_____	_____
f. Calcium chloride	_____	_____	_____	_____	_____
g. Ammonium chloride	_____	_____	_____	_____	_____
h. Silver chloride	_____	_____	_____	_____	_____
i. Zinc oxide	_____	_____	_____	_____	_____
j. Calcium oxide	_____	_____	_____	_____	_____

Experiment 6   Report Sheet   55

		Location in Well Plate	Water sol?	Conductivity?	Acid sol?	Base sol?
k.	Silicon dioxide	_____	_____	_____	_____	_____
l.	Sodium carbonate	_____	_____	_____	_____	_____
m.	Potassium carbonate	_____	_____	_____	_____	_____
n.	Calcium carbonate	_____	_____	_____	_____	_____
o.	Calcium sulfate	_____	_____	_____	_____	_____
p.	Copper sulfate pentahydrate	_____	_____	_____	_____	_____
q.	Potassium permanganate	_____	_____	_____	_____	_____
r.	Aluminum powder	_____	_____	_____	_____	_____
s.	Zinc powder	_____	_____	_____	_____	_____
t.	Tin powder	_____	_____	_____	_____	_____
u.	Carbon, as charcoal	_____	_____	_____	_____	_____
v.	Magnesium	_____	_____	_____	_____	_____
w.	Iron	_____	_____	_____	_____	_____
x.	Sucrose	_____	_____	_____	_____	_____

Part 2

Substance		Acetone sol?	Water sol?	Conductivity	Observations
a.	Methanol	_____	_____	_____	_____
b.	2-Propanol	_____	_____	_____	_____
c.	Ethylene Glycol	_____	_____	_____	_____
d.	Cholesterol	_____	_____	_____	_____
e.	Naphthalene	_____	_____	_____	_____
f.	p-Dichlorobenzene	_____	_____	_____	_____
g.	Acetic acid	_____	_____	_____	_____

56 Experiment 6 Report Sheet

(cont'd)	Acetone sol?	Water sol?	Conductivity	Observations
h. Polystyrene				
i. Benzoic acid				
j. Sucrose				
k. Caffeine				
l. Citric acid				
m. Cyclohexane				
n. *tert*–Butyl methyl ether				
o. Ethyl acetate				
p. Iodine				
q. Vanillin				
r. Acetone				

POSTLABORATORY QUESTIONS

1. Write down the chemical structures of the substances in Part 2 of this experiment. Look up the substances in an organic chemistry textbook or the *Aldrich Catalog Handbook of Fine Chemicals*. Some of these structures are quite complex.

2. Can you make any generalizations regarding the relationship between the structures of the substances you have tested and their solubilities and conductivities?

EXPERIMENT 7

Isolation of the Silver in a Dime

Introduction

At one time all the currency in the United States was backed by precious metals. One could go to the bank and trade in a piece of paper money for an equivalent quantity of gold or silver. In 1932 this country went off the gold standard, but for many years paper currency was marked "Silver Certificate," meaning that one could demand silver dollars for the paper bill. Prior to 1965, dimes, quarters, and half-dollars were made of a copper/silver alloy with the silver predominating. The copper was added to make the silver harder and, therefore, the money more durable. It was called "coin silver" to distinguish it from "sterling silver," which is 92.5% silver and 7.5% copper.

The object of this experiment is to isolate pure silver from a pre-1965 dime and to determine the percent silver in the original dime. By consulting a newspaper it is possible to find the current selling price for pure silver, so the present-day value of a dime can be calculated. Pre-1965 dimes can be obtained from coin dealers; the silver dime is not much more expensive than many of the reagents we use in the laboratory.

Silver is not attacked by nonoxidizing acids, but it is readily attacked by nitric acid. The silver is oxidized by the acid and the nitric acid is reduced to nitrogen oxide or nitrogen dioxide. No hydrogen is given off. When a dime is dissolved in nitric acid the following reactions can occur:

$$Cu(s) + 4\ HNO_3\ conc\ (aq) \rightarrow Cu(NO_3)_2(aq) + 2\ NO_2(g) + 2\ H_2O$$

$$3\ Cu(s) + 8\ HNO_3\ dil\ (aq) \rightarrow 3\ Cu(NO_3)_2(aq) + 2\ NO(g) + 4\ H_2O$$

$$Ag(s) + 2\ HNO_3\ conc\ (aq) \rightarrow AgNO_3(aq) + NO_2(g) + H_2O$$

$$3\ Ag(s) + 4\ HNO_3\ dil\ (aq) \rightarrow 3\ AgNO_3(aq) + NO(g) + 2\ H_2O$$

It is difficult to determine whether colorless NO is given off in the reaction because it is immediately oxidized to the red-brown gas, NO_2. With continued evolution the gas in the tube can appear colorless, but will turn brown at the mouth of the tube as it reacts with oxygen.

$$2\ NO(g) + O_2(g) \rightarrow 2\ NO_2(g)$$

After the reaction is complete the acidic solution contains silver ions, Ag^+, and copper ions, Cu^{2+}. To separate these two ions we take advantage of the fact that silver chloride is almost insoluble and copper(II) chloride is very soluble in aqueous solution. So by simply adding a soluble halide (potassium chloride, bromide, or iodide) we obtain a precipitate of insoluble silver halide that can be separated from the copper ion solution by filtration. In this experiment we will use the chloride salt.

When the silver chloride is first precipitated the particle size is very small, and thus it is difficult to isolate by filtration, but if the fine silver chloride is acidified with a drop of 6 M

58   Experiment 7   Isolation of the Silver in a Dime

nitric acid and boiled it will coagulate into much coarser pieces that are easy to collect. This coarse precipitate is collected by filtration, washed free of copper ion, and then dried.

To return the silver chloride to metallic silver it must be *reduced*. This is the basis of most photography: Silver halide exposed to light can be reduced to black metallic silver (a photographic negative) under the proper conditions. In the present experiment ordinary sucrose will be employed as the reducing agent. It, in turn, is oxidized to a carboxylic acid. The silver obtained at this point is a coarse powder. Using a gas/oxygen torch, it can, in an optional experiment, be melted into a globule of familiar-looking silver metal.

From both the weights of the dry silver chloride and the dry metallic silver you will calculate the percent silver in a dime.

Procedure Summary

A dime is dissolved in nitric acid, the silver precipitated as the chloride which in turn is reduced to metallic silver. The weight of the silver compared to the dime allows calculation of the percent silver in the dime.

Prelaboratory Assignment

Read the Introduction and Procedure sections and answer the Prelaboratory Questions on the Report Sheet.

Materials

Apparatus

Milligram balance
50-mL beaker
2-in. watch glass
Graduated cylinder
10-mL Erlenmeyer flask
Stirring rod
Steam bath or hot plate adjusted to 95°C

Reagents

Pre-1965 dime for five students
15 mL nitric acid, concentrated, for five students
2 M potassium chloride solution, about 4 mL
Dilute nitric acid (5 mL 6 M nitric acid in 100 mL water)

15 mL acetone
4 mL 6 *M* sodium hydroxide solution
1 g sucrose

Safety Information

1. **Safety glasses must be worn at all times in the laboratory.**
2. **Handle concentrated nitric acid with care.**
3. **Acetone is flammable.** Extinguish all flames when using acetone.
4. **Nitrogen dioxide is toxic.** Dissolve dime in the hood.

Procedure

A group of five students should perform the first part of this experiment:

1. To a tared (previously weighed) 50-mL beaker add a pre-1965 dime and determine, to the nearest milligram, the weight of the dime. Add 5 mL of distilled water and 15 mL of concentrated nitric acid to the beaker in the hood and cover it with a watchglass. Note the appearance of this reaction, the gas that evolves, and the color of the solution. If the reaction does not proceed rapidly enough, warm the beaker on a steam bath. A very vigorous reaction will ensue, at which point heating is no longer needed. If crystals of silver nitrate form in the beaker add a small quantity of water to dissolve them.

2. Weigh the beaker and its contents on the milligram balance. Then divide the resulting solution into five parts by weighing it (more accurate than measuring it volumetrically) into four tared 10-mL Erlenmeyer flasks. Each flask need not contain exactly one-fifth of the solution, but the weight of each solution should be recorded to the milligram, so that the fraction of the original solution can be calculated.

3. Each student should add 2 *M* potassium chloride solution dropwise to the one-fifth aliquot (portion) of the original solution in the Erlenmeyer flask. Note the formation of a precipitate of silver chloride. Continue adding the chloride ion solution dropwise until no more precipitate forms. A small excess of the chloride is desirable to make sure all the silver ion is precipitated, but avoid a large excess, which can redissolve some of the precipitate.

    On a steam bath, or an electric heater adjusted to just under the boiling point of water, heat the solution almost to boiling for a few minutes to coagulate the precipitate. If the solution is boiled it will often bump out of the flask as steam forms between the precipitate and the bottom of the flask.

4. Remove the flask from the heat and allow the precipitate to settle. Stirring will increase the rate of coagulation. Decant (pour off) the supernatent (the clear aqueous layer above the precipitate). Do not, of course, lose any of the precipitate. Wash the precipitate with four 5-mL portions of dilute nitric acid (5 mL of 6 *M* acid in 100 mL of water) while it is

in the flask (Figure 7.1). Stir the precipitate with a glass rod and break up the lumps. Each time, allow the precipitate to settle and pour off as much of the liquid as possible without losing any of the silver chloride. If necessary wash the stirring rod.

Figure 7.1 Reagent bottle with 1-mL graduated Beral pipette.

5. To about 2 mL of the last wash liquid add about 0.5 mL of concentrated ammonium hydroxide. If copper ion is present the solution will turn dark blue due to the formation of the tetraammine copper complex, and the precipitate must be rinsed at least once more:

$$\underset{\text{light blue}}{Cu^{2+}} + 4\ NH_4OH \rightarrow \underset{\text{dark blue}}{[Cu(NH_3)_4]^{2+}} + 4\ H_2O$$

6. After all the copper ion has been removed from the precipitate by washing with water, wash the precipitate with three 3-mL portions of acetone. Acetone is miscible (uniformly mixable) with water, but has a much lower boiling point and is therefore much easier to evaporate from the silver chloride. Pour off as much of the acetone as possible and place the flask in a labeled beaker in a warm place so that the acetone, which boils at 56°C, can evaporate completely. Warm the flask until its weight is constant, which will indicate all of the acetone has evaporated. But remember the flask must be cool in order to weigh it.

7. Add to the dry silver chloride 4 mL of distilled water, 3 mL of 6 M sodium hydroxide solution, and 1 g of sucrose (table sugar). Heat the mixture to near boiling on the hot plate or steam bath and stir it with a glass rod until the silver chloride is entirely converted to gray metallic silver. Decant the aqueous phase and wash the silver four times with 5-mL portions of warm water. Then, after pouring off as much water as possible, wash the silver with three 3-mL portions of acetone, and dry the flask and contents to constant weight as before.

8. To prepare a silver button (optional) mix the silver powder with some sodium carbonate, which serves as a flux and prevents the silver particles from flying around. Place the mixture in a small depression hollowed out of a block of charcoal and heat the metal with a gas-oxygen torch until the metal is molten. After the button cools, wash it with dilute nitric acid to remove any sodium carbonate.

Cleaning Up

Collect all the aqueous material in one big beaker and neutralize it with solid sodium bicarbonate (baking soda). Flush the neutral solution down the drain. It contains a small quantity of copper ion.

In many jurisdictions the acetone wash liquid can be washed down the drain with water also. It can also be collected as flammable waste in the appropriate container.

Name _____ Section _____

Lab Instructor _____ Date _____

EXPERIMENT 7 Isolation of the Silver in a Dime

PRELABORATORY QUESTIONS

1. Why does silver chloride turn dark on standing?

2. In an analysis of the silver in a dime a student obtained 3.696 g of silver bromide. What weight of silver can be obtained from this silver bromide theoretically? If 2.000 g of silver was actually obtained, what is the percent yield of silver?

DATA AND CALCULATIONS

Mass of beaker	_____
Mass of dime	_____
Mass of beaker, dissolved dime, and aqueous solution	_____
Mass of solution	_____
Mass of Erlenmeyer flask plus about one-fifth of dime solution	_____
Mass of Erlenmeyer flask	_____
Mass of solution	_____
Fraction of entire solution in flask	_____
Mass of flask plus silver chloride	_____
Mass of flask	_____

Experiment 7 Report Sheet

Mass of silver chloride _____

Atomic weight of silver _____

Molecular weight of silver chloride _____

Fraction of silver in silver chloride _____

Calculated mass of silver
in your silver chloride _____

Calculated mass of silver
in your dime from weight of AgCl _____

Percent of silver in dime from AgCl weight _____

Mass of silver obtained
by reduction of AgCl _____

Calculated mass of silver in dime
from weight of silver in aliquot _____

Percent of silver in dime from silver mass _____

Cost of dime from coin dealer _____

Current price of silver per gram (The price is usually given in Troy ounces; one Troy oz = 31.04 g.) _____

Value of the silver in your dime _____

POSTLABORATORY QUESTIONS

1. What are the principal causes of experimental error in this experiment?

2. In the reaction equations given for the oxidation of silver by nitric acid (both concentrated and dilute), identify both the oxidizing and reducing agents.

EXPERIMENT 8

Synthesis of Manganese(II) Chloride

Introduction

In this experiment you will synthesize manganese(II) chloride by allowing a known mass of manganese metal to react with an excess of hydrochloric acid. From the theoretical yield (calculated from the mass of metal used) and the actual yield, you will be able to determine the percentage yield. The equation for the reaction is

$$Mn(s) + 2\ HCl(aq) \rightarrow MnCl_2(aq) + H_2(g)$$

Procedure Summary

A sample of manganese is dissolved in hydrochloric acid to produce manganese(II) chloride.

Prelaboratory Assignment

Read the Introduction and Procedure sections carefully and answer the Prelaboratory Question on the Report Sheet.

Materials

Apparatus

Electrically heated sand bath
10-mL Erlenmeyer flask or small (10 × 75-mm) test tube or culture tube
Desiccator
Forceps, for handling hot flask

Reagents

Manganese metal
10 M hydrochloric acid

Safety Information

1. **CAUTION! Ten molar hydrochloric acid is strongly corrosive.** It will cause severe burns to the eyes and skin. Wear safety glasses and an apron when handling. Use only as directed and do not remove it from the hood. Wash spills from skin and other surfaces immediately. During the evaporation step, fumes of the acid are evolved. This is especially hazardous to contact lens wearers. Evaporation is conducted on a hot sand bath. Carry this out in a hood or with adequate ventilation. Use care to avoid burns because the sand is very hot.

Procedure

1. Determine the mass of a 10-mL Erlenmeyer flask or labeled test tube to the nearest milligram. Place about 0.05 to 0.10 g of manganese metal in the flask and determine the mass of flask and contents again. In the fume hood, carefully add 30 drops of 10 M hydrochloric acid, using the pipette provided. Replace the pipette in the tube attached to the hydrochloric acid bottle (see Figure 7.1 in Experiment 7). Allow the acid to react with the metal until the sample has dissolved entirely (about 5 minutes).

   During dissolution, determine whether the process is exothermic or endothermic. Place the flask with the dissolved sample in a heated sand bath at an angle, as demonstrated by your instructor. As the contents of the flask near dryness, tilt it with the mouth of the flask on the rim of the sand bath. Allow the flask to heat to dryness, rotating it occasionally. Your product should be a pale, pinkish-white solid. When the flask is dry, remove it *carefully* from the sand bath and allow it to cool for at least 5 minutes in a desiccator. If time is short, this is a good stopping place.

2. Remove the cooled flask from the desiccator and determine the mass of the flask and product. *Do not weigh a hot flask.* You will get poor results and you may damage the balance. Return the flask to the hot sand bath for an additional 3 to 5 minutes of heating. As before, lean the flask on its side, rotating it periodically. Again, cool the flask in a desiccator and reweigh the flask and contents. This mass should agree to within 3 mg (±0.003 g) of the previous value. If the mass has changed by more than 5 mg repeat the heating and cooling steps until you arrive at a constant mass.

Cleaning Up

Label your tube and hand it in to your instructor. Manganese ion is toxic to the environment. The manganese(II) chloride requires special handling for disposal, so do *not* wash it down the sink.

Name _____   Section _____

Lab Instructor _____   Date _____

EXPERIMENT 8   Synthesis of Manganese(II) Chloride

PRELABORATORY QUESTION

1. If you knew the atomic weight of manganese and of chlorine could you use this experiment to determine the number of chlorine atoms bound to manganese in the product? Explain you answer.

DATA AND CALCULATIONS

Mass of flask   _____

Mass of flask plus manganese   _____

Mass of manganese used   _____

Moles of manganese used   _____

Mass of flask   _____

Mass of flask plus manganese chloride   _____

Mass of manganese chloride prepared   _____

Moles of manganese chloride prepared   _____

Theoretical yield of $MnCl_2$ in grams   _____

Your percent yield of $MnCl_2$ _____

CONCLUSIONS

1. Explain the effect on your percent yield of each of the following errors:

    a. Most or all of the acid solution is evaporated.

Experiment 8   Report Sheet

b. Heating is continued too long, or at too high a temperature, and some of the manganese(II) chloride decomposes.

c. While cooling, the flask with manganese chloride product is left to cool in open air, rather than in a desiccator.

2. Given the possible errors suggested above, as well as any others you can think of, is a student who starts with pure, high-quality reagents more likely to report greater or less than 100% yield? Explain your reasoning.

3. Suppose you had spilled a small amount of the 10 M HCl. Describe in step-by-step fashion what you should do; do not leave any steps out.

4. Suppose sulfuric acid, H_2SO_4, is substituted for HCl
   a. Write a balanced equation for the reaction between sulfuric acid and manganese metal. Assume that hydrogen gas is one of two products formed.

   b. Explain the changes in calculations that would be necessary.

EXPERIMENT

9 Analysis of Copper Oxide

Introduction

The purpose of this investigation is to determine whether the copper in a sample of copper oxide is copper(I) (Cu_2O) or copper(II) (CuO). To solve this problem, a sample of the oxide is dissolved in hydrochloric acid, forming a solution of copper(I or II) chloride (CuCl or $CuCl_2$) in the process. This solution will then be allowed to react with metallic aluminum, resulting in formation of aqueous aluminum chloride and the precipitation of metallic copper.

$$Cu_2O + 2\ HCl \rightarrow 2\ CuCl + H_2O$$

$$3\ CuCl + Al \rightarrow AlCl_3 + 3\ Cu$$

or

$$CuO + 2\ HCl \rightarrow CuCl_2 + HCl$$

$$3\ CuCl_2 + 2\ Al \rightarrow 2\ AlCl_3 + 3\ Cu$$

The reaction of aluminum metal with copper ion to give copper metal and aluminum ion is an example of an oxidation–reduction reaction, called for short, a *redox* reaction. In this reaction the copper ion is reduced to metallic copper and the aluminum metal is oxidized. This equation can be split into two half reactions, one showing the reduction and the other the oxidation:

$$Cu^+\ e^- \rightarrow Cu\ \text{reduction of copper (I) ion}$$

$$Al \rightarrow Al^{3+} + 3\ e^-$$

When the copper is dried and weighed, the mass of oxygen in the original sample of copper oxide may be determined by difference, and the empirical formula of the copper oxide can then be calculated.

Procedure Summary

An unknown copper oxide is reacted with hydrochloric acid to give a solution of copper ion. This copper ion is then allowed to react with aluminum metal to give a precipitate of copper metal. This precipitated metal is carefully dried and weighed. From the weight of the copper metal formed and the weight of the copper oxide used, it is possible to calculate whether the oxide was copper(I) or copper(II) oxide.

Experiment 9 Analysis of Copper Oxide

Prelaboratory Assignment

Read the Introduction and Procedure sections carefully and answer the Prelaboratory Question on the Report Sheet.

Materials

Apparatus

20 - or 30mL beaker
Pasteur pipette or thin-stem Beral pipette

Reagents

Copper(I) or copper(II) oxide unknowns
6 M hydrochloric acid, 2.5 mL
Deionized or distilled water
Aluminum metal, strip or wire, about 1 g

Safety Information

1. **Safety goggles must be worn at all times in the laboratory.**
2. **Handle hydrochloric acid with great care.** All spills of hydrochloric acid must be cleaned up with lots of water and reported to your teacher.

Procedure

1. Label and weigh (±0.001 g) a clean, dry 20 - or 30mL beaker, add 0.25 to 0.35 g of the unknown copper oxide from the supply bottle and then reweigh the beaker and contents. Using the graduated 1-mL plastic Beral pipette in the test tube attached to the reagent bottle, carefully add about 2.5 mL of 6 M hydrochloric acid. Gently swirl the beaker to ensure that the acid makes good contact with all of the copper oxide.

Caution! Do not get your face over the beaker and do not breathe the fumes of hydrochloric acid.

---

2. While waiting for all of the copper oxide to dissolve, polish a strip of aluminum or an aluminum wire with fine steel wool to remove the oxide coating. Wipe the strip or wire clean with a cloth or paper towel and wrap it into a loose coil. When the unknown copper oxide has been completely digested by the acid, carefully add about 5 mL of deionized or distilled water, a little at a time, with constant swirling, to the solution. Once this dilution is complete, add the loosely coiled strip of aluminum metal to the contents of the flask. Observe the reaction that takes place between the dissolved copper ions and the metallic aluminum. How can you determine when the reaction is complete?

3. When the reaction is complete, remove the unreacted aluminum, using a glass rod, and place it in a beaker. Clean up any drips, then carefully decant (pour off) into another container as much of the liquid as you can without loss of solid product. Add about 5 mL of deionized or distilled water, swirl, and decant again. This time, use a Beral or Pasteur pipette to remove as much of the water as possible. Do not lose any of the copper in this process.

4. Once again add about 5 mL of either deionized or distilled water to your product. Decant as much water as possible, then use the pipette to remove the last of the liquid. Add 2 mL of acetone to the flask and mix it thoroughly, allowing the copper to settle. Remove the acetone with a Beral or Pasteur pipette using great care not to remove any of the copper. Repeat this process with another 2 mL of acetone, heat it on a steam bath to drive off most of the acetone and then set the flask in a warm oven to allow the last traces of the acetone to evaporate. After the flask is cool weigh it carefully.

Cleaning Up

The combined rinsings and filtrate are safe to be washed down the drain without further treatment. The copper can be disposed of in the waste basket or it can be, at the discretion of your instructor, reconverted to copper oxide.

Name _____ Section _____

Lab Instructor _____ Date _____

EXPERIMENT 9 Analysis of Copper Oxide

PRELABORATORY QUESTION

1. Write balanced equations for the reaction of CuO with HCl and the subsequent reaction with Al. Calculate the percent by weight of copper in CuO and Cu_2O.

DATA AND CALCULATIONS

1. Mass of Erlenmeyer flask   _____

2. Mass of Erlenmeyer flask plus copper oxide   _____

3. Mass of copper oxide unknown   _____

4. Amount of 6 M hydrochloric acid added   _____

5. Mass of Erlenmeyer flask   _____

6. Mass of Erlenmeyer flask plus copper   _____

7. Mass of copper   _____

CONCLUSIONS

1. Use the mass of your original copper oxide sample and the mass of copper recovered to calculate the mass percentage of copper in the oxide. Compare your result with the theoretical values you calculated in the Prelaboratory Assignment. What do you conclude is the formula of the starting material?

2. Give the proper name for the copper oxide with which you began. Assuming the copper did not change oxidation number as a result of having dissolved in acid, what is the name and what is the formula for the copper chloride you produced when the copper dissolved in the hydrochloric acid?

POSTLABORATORY QUESTIONS

1. What volume of 6 M hydrochloric acid is required theoretically to react with 0.30 g of copper(I) oxide?

2. How did you know the reaction between copper ion and aluminum was complete?

3. a. Calculate the approximate number of moles of hydrochloric acid used in this experiment.

   b. What are the possible ratios of moles of hydrochloric acid to moles of copper oxide, assuming the mass of oxide used is 0.30g and the volume of 6 M HCl is 2.5 mL. Answer for both copper (I) oxide and copper (II) oxide.

   c. Which is the limiting reagent in each case, the oxide or the acid?

EXPERIMENT

10 Gravimetric Determination of Chloride Ion Concentration

Introduction

The chloride ion concentration in a sample of an unknown solid or solution can be determined by precipitating the chloride as silver chloride. This is done by addition of an excess of silver nitrate, forming a precipitate that is easy to filter, wash, and dry in micro- or semi-microquantities. The mass of the precipitate is used to calculate stoichiometrically the amount of chloride ion present in the unknown. This experiment is an example of *gravimetric analysis*. The mass percent of chloride is determined *gravimetrically*. In this investigation, you will start with a mixture of barium chloride and potassium chloride; your goal is to determine the mass percent of each component in the mixture.

Procedure Summary

A filter pipette is prepared and then silver nitrate solution is added to a solution of the unknown mixture of barium and potassium chlorides. The precipitate of silver chloride is removed by filtration, washed thoroughly, and dried in an oven. From the weight of the silver chloride—and thus the moles of silver chloride—the number of moles of chloride ion in the unknown can be calculated. With this knowledge the percents by weight of the two components of the mixture can be calculated.

Prelaboratory Assignment

Read the Introduction and Procedure sections carefully and answer the Prelaboratory Questions on the Report Sheet.

Materials

Apparatus

Milligram or analytical balance
Pasteur pipette
Cotton
Wood stick (applicator stick)
10 × 75-mm culture tube
1-mL volumetric pipette, calibrated pipette, or syringe with Eppendorf tip (for solution unknowns only)
Centrifuge
Drying oven (105-110°C)

Reagents

8 M nitric acid
distilled water, in wash bottles
1.00 M silver nitrate solution, $AgNO_3$
Acetone
Unknowns

Safety Information

1. **Safety goggles must be worn at all times in the laboratory.**
2. **Nitric acid is extremely dangerous. If you spill it anywhere, use large volumes of water to wash it up.**
3. **Silver nitrate is harmful and will stain skin and clothing.** Wash spills with large amounts of water. If it gets on your clothes or skin, ask your instructor about removal.
4. **Acetone is highly flammable.** Be sure there are no open flames in the laboratory before proceeding.

Procedure

Your instructor will demonstrate preparation of the filter pipette. Unknowns may be solids or solutions.

Experiment 10   Gravimetric Determination of Chloride Ion Concentration   77

Part 1: Preparation of the Filter Pipette

1. Remove and discard the extended tip from a Pasteur pipette. Plug the bottom (narrow) end of the pipette with dry cotton (not fiberglass or glass wool) using a wooden stick to pack the cotton firmly in place (Figure 10.1). Place the packed pipette in a 105 to 110°C oven. While it is drying complete preparation of the sample through the rinsings (Step 6).

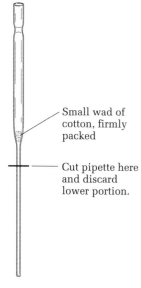

Figure 10.1   Filter pipette.

Part 2: Preparation of the Sample

Solid Unknowns

2. Weigh a clean, dry 10×75-mm culture tube. Transfer a small amount (about the size of a single grain of rice) of the unknown to the tube and determine the mass of the sample. The sample size should be between 50 and 120 mg if an analytical balance (± 0.1 mg) is available. If the balance being used is limited to milligram sensitivity (± 1 mg) the sample mass should be in the 100- to 150-mg range. Dissolve the crystals in about 1 mL of distilled (deionized) water. Warming the tube (water bath or sand bath) may help if the solid is slow to dissolve. If it is necessary to add more water, do not exceed one-third of the tube's capacity. Once all of the sample is in solution, proceed to Part 3 or for solution unknowns use the following procedure.

Solution Unknowns   For solutions, the mass of unknown mixture used to prepare a known volume (e.g., 100.0 mL) of solution is provided; use this and your results to determine the percent by weight of each component in the mixture.

3. Place exactly 1.00 mL of the unknown solution in a clean, dry 100-mm reaction tube. Use either a volumetric pipette, a calibrated pipette, or a syringe fitted with an Eppendorf tip; a graduated cylinder is not sufficiently precise.

Part 3: Precipitation of Silver Chloride

4. Add 2 drops of 8 M HNO_3 to the solution in the tube and shake gently. Add about 1 mL of 1.00 M $AgNO_3(aq)$, dropwise and with gentle shaking to the solution in your tube.

5. When the filter pipette from Step 1 has been in the oven for at least 15 minutes, remove it to a desiccator to cool. Centrifuge your tube to compact the precipitate, then test the supernatant for unprecipitated chloride ion by adding 1 drop of 1 M $AgNO_3(aq)$. If cloudiness is observed, add 5 more drops of silver nitrate solution, mix, and centrifuge again. If no cloudiness appears, decant the clear supernatant from the test tube, taking care not to lose any of the precipitated silver chloride.

6. To rinse the precipitate, add distilled (or deionized) water dropwise, shaking the tube so as to disperse the solid. Centrifuge the tube (Figure 10.2), then decant and discard the liquid portion. Repeat this rinsing twice more, but do not decant the final rinsing.

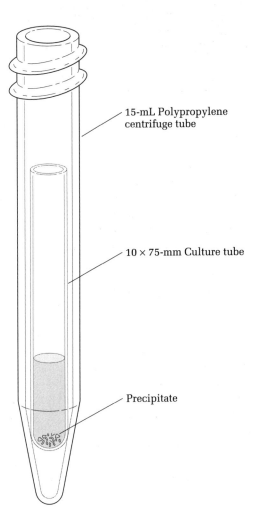

Figure 10.2 Apparatus for centrifugation of precipitate.

Part 4: Filtration of the Silver Chloride Precipitate

7. Remove the cooled filter pipette from the desiccator and determine its mass. Use a piece of heavy-walled rubber tubing to connect the packed (narrow) end of the filter pipette to an aspirator. With the aspirator running, place the wide end of the pipette into the tube containing the precipitate and final rinsing, and draw the liquid and precipitate up into the cotton packing (Figure 10.3). Without removing the pipette from the tube, use a wash bottle to rinse all of the solid from the tube up into the pipette.

8. With the pipette still in the test tube, add about 0.5 mL of acetone to the tube and draw it up through the pipette; repeat. Allow the aspirator to continue to pull air through the filter pipette for about 1 minute to help dry the precipitate. Place the pipette in the drying oven for 10 to 15 minutes, then remove it to the desiccator and allow it to cool for 15 minutes (longer if an analytical balance is used). Weigh filter pipette and contents.

9. Return the pipette to the oven for an additional 15 minutes, then allow it to cool in the desiccator, and reweigh. If the weights do not agree within experimental error (±0.5% of the precipitate mass), continue the alternate heating and cooling until agreement is obtained.

Experiment 10   Gravimetric Determination of Chloride Ion Concentration   79

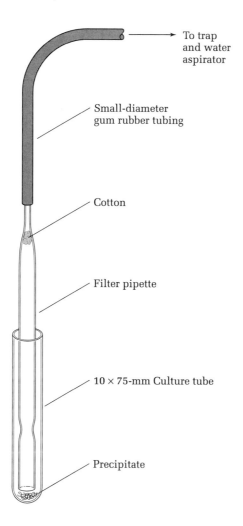

Figure 10.3 Collection of precipitate on filter pipette.

Cleaning Up

Place the the filter pipette, which contains a small amount of silver chloride, in the container provided.

Name _____ Section _____

Lab Instructor _____ Date _____

EXPERIMENT 10   Gravimetric Determination of Chloride Ion Concentration

PRELABORATORY QUESTIONS

1. Write the balanced net-ionic equation for precipitation of chloride ion by silver ion in aqueous solution.

2. If 100 mg of $BaCl_2 \cdot 2\, H_2O$ is dissolved in water and then an excess of 1.00 M $AgNO_3(aq)$ is added, calculate the mass of silver chloride that will form, assuming complete precipitation.

3. What volume of 1.00 M silver nitrate solution would be needed to completely precipitate the chloride from the barium chloride in Question 2?

4. Why is it important to use an excess of silver nitrate? If you knew the atomic weight of manganese and of chlorine, could you use this experiment to determine the number of chloride ions bound to manganese in the product of Experiment 8? Explain your answer.

DATA AND CALCULATIONS

(Solid unknowns only)
Mass of sample vial before removal of sample _____

Mass of sample vial after removal of sample _____

Mass of sample used _____

81

Mass of pipette and precipitate _____

Mass of packed filter pipette
before collection of precipitate _____

Mass of silver chloride precipitate _____

Moles of silver chloride precipitate _____

If your unknown was a solution, calculate the total number of moles of chloride in the stock solution _____

CONCLUSIONS

1. The number of moles of silver chloride must equal the number of moles of chloride ion in your sample. Why?

2. Further, the moles of chloride ion must equal the sum of the moles of potassium chloride in the sample, plus twice the number of moles of barium chloride. Why?

   Represent this total by the letter t:

   $$\text{mol Cl}^- = \text{mol KCl} + 2\,(\text{mol BaCl}_2) = t$$

3. Further, if we let n represent the number of moles of chloride that originated with the potassium chloride, then $t - n$ gives the number of moles of chloride originally in the barium chloride. From this, we can deduce that the mass of the original mixture sample, m, is given by

   $$n(\text{MW of KCl}) + \frac{(t-n)}{2}(\text{MW of BaCl}_2) = m$$

   Carry out the necessary calculations to determine the composition of the original mixture. Show your work.

POSTLABORATORY QUESTIONS

1. What would be the effect on your chloride percentage if:

   a. insufficient silver nitrate solution were used? Explain.

b. drying of the filter and precipitate was not complete? Explain.

c. the filter pipette and precipitate was not completely cooled before weighing? Explain.

2. Conclude your report with a discussion of experimental errors and identify those portions of the procedure where extra care is needed to ensure satisfactory results. Include, but do not limit your analysis to, the answers you gave in the preceding question. Consider only those factors that might apply to your own experiment.

3. The purpose of adding the nitric acid during sample preparation is to remove traces of carbonate ion, thus preventing coprecipitation of silver carbonate along with the silver chloride. In Experiment 11 (gravimetric sulfate determination), hydrochloric acid will be used for the same purpose. Why is hydrochloric acid not suitable in this case?

4. A sample of a pure unknown chloride (0.0212 g) gave 0.0716 g of silver chloride. Identify the unknown chloride.

EXPERIMENT 11

Gravimetric Determination of Sulfate Ion Content

Introduction

In this experiment you will be given a sample that contains sulfate ion. It could be a pure substance, such as sodium sulfate, or it could be a mixture of some sulfate-containing substance such as nickel sulfate mixed with an "inert" substance such as sodium chloride. Your task is to determine the mass percent of sulfate ion in the original sample. This is an exercise in careful technique. You will be evaluated on how close you come to the "correct" answer.

The sulfate-containing unknown will dissolve in water. When an aqueous solution of barium chloride is added to the solution, insoluble barium sulfate will precipitate from the solution:

$$SO_4^{2-}(aq) + Ba^{2+}(aq) \rightarrow BaSO_4(s)$$

This precipitate is collected by centrifugation, washed carefully, dried, and weighed. From the weight of the barium sulfate collected, the number of moles of barium sulfate can be calculated, and thus the number of moles of sulfate ion, which can be used to calculate the weight of sulfate in the unknown. Knowing the weight of the original unknown and the weight of the sulfate ion then allows the percentage of sulfate ion in the unknown to be calculated. This experiment is another example of *gravimetric analysis*. The percent of sulfate ion in the unknown is determined *gravimetrically*. You should plan on spending about 1 hour on this experiment the first day, plus short periods of time on one or more successive days—perhaps while another experiment is in progress.

Procedure Summary

A very small sample of an unknown sulfate compound is dissolved in water and the sulfate precipitated as barium sulfate. If we know the formula weight of barium sulfate, we can calculate the number of moles of sulfate ion and then the weight of sulfate ion in the unknown, and thus determine the percent sulfate in the unknown.

Prelaboratory Assignment

Read the Introduction and Procedure sections carefully and answer the Prelaboratory Questions on the Report Sheet.

Materials

Apparatus

Milligram or analytical balance
100-mm test tube
Electrically heated sand bath and boiling water bath
Desiccator
Centrifuge

Reagents

1 M HCl
1.0 M BaCl$_2$

Safety Information

1. **Safety goggles must be worn at all times in the laboratory.**
2. **Barium chloride is toxic by ingestion.** Wash your hands before leaving the laboratory.

Procedure

For best results, duplicate determinations should be carried out. The sample mass should be in the range of 100 to 150 mg if a balance with milligram sensitivity is used, or 50 to 100 mg if an analytical balance is available. Label both samples clearly for ease of identification.

Part 1: Sample Preparation

1. Determine the mass of a clean, dry 100-mm reaction tube. Place a small amount of the unknown (about the volume equivalent of one grain of rice) in the tube, then reweigh the tube and contents. Add about 1 mL of distilled (or deionized) water to dissolve the sample, warming the tube in a hot water bath or on a sand bath if necessary. Some solids may require more water, but be careful not to fill the tube beyond about one-third full. The sample must be completely dissolved before proceeding.

Part 2: Precipitation of Barium Sulfate

2. Add 2 drops of 1 M HCl to the tube and shake gently. Follow this with 0.5 to 1.0 mL of 1.0 M BaCl$_2$ added dropwise, with gentle shaking after each 2 to 3 drops. Take care not to get barium chloride on your skin; if you do, wash it off with lots of water. Place the

tube and contents in a boiling water bath for 3 to 5 minutes, or on a hot sand bath for 1 to 2 minutes, to facilitate coagulation of the precipitate. Remove the tube, allow it to cool, then centrifuge for 30 seconds.

3. Without disturbing the solid on the bottom of the test tube, add 1 more drop of the barium chloride solution. If no new cloudiness appears, proceed with the washing of the precipitate. If cloudiness is noted, add 5 more drops of the $BaCl_2$, then heat, centrifuge, and test again with barium chloride; continue in this fashion until addition of barium chloride does not cause further cloudiness.

Part 3: Washing the Precipitate

4. Decant and discard the clear supernatant solution above the barium sulfate precipitate, being careful not to lose any solid. Add about 10 drops of distilled (deionized) water, then shake the tube and contents until all the precipitate is suspended in the water (it will not dissolve). Centrifuge the suspension and again discard the clear, colorless supernatant, being careful not to lose any of the white solid. Repeat the rinsing, centrifuging, and decanting twice more, followed by a final rinsing with acetone.

5. Dry the test tube and precipitate first on a steam bath to drive off most of the acetone and then in a 105 to 110°C oven for at least 1 hour, preferably overnight.

Part 4: Weighing the Precipitate

6. Remove the tube from the oven and allow it to cool for a minimum of 30 minutes in a desiccator. Determine the mass of the tube and contents. Return the tube to the oven for at least 1 hour, cool in the desiccator once more, then reweigh it. If the mass agrees within experimental error (±0.5% of the precipitate mass) with the previous value, the experiment is completed. If not, continue the cycle of oven-drying, cooling, and weighing until a constant mass is obtained.

Cleaning Up

Barium sulfate, like barium chloride, is toxic by ingestion. Depending on local regulations, you may be directed to transfer your solid product to an appropriate container for removal as hazardous waste. Small amounts of barium ion were washed into the effluent stream during the decanting and rinsing portions of the procedure. This is unavoidable, but the amounts are below the parts-per-trillion level, that is, so small as to present no significant hazard to the environment.

Name _____   Section _____

Lab Instructor _____   Date _____

EXPERIMENT 11   Gravimetric Determination of Sulfate Ion Content

PRELABORATORY QUESTIONS

1. Write the balanced, net-ionic equation for the reaction between barium and sulfate ions in aqueous solution to form barium sulfate.

2. Suggest a reason why it is desirable to use an excess of barium ion in the precipitation.

3. A sample of an unknown sulfate compound has a mass of 0.1000 g. Addition of excess barium chloride solution forms a barium sulfate precipitate of mass 0.0676 g. What is the mass percent sulfate ion in the unknown compound?

4. Hydrated calcium sulfate, $CaSO_4$, contains 55.8% sulfate by mass. Calculate the number of waters of hydration present in the hydrate.

DATA AND CALCULATIONS

Unknown number _____

	Trial 1	Trial 2
Mass of empty tube	_____	_____
Mass of tube plus unknown	_____	_____
Mass of unknown	_____	_____
Mass of tube plus $BaSO_4$	_____	_____
Mass of empty tube	_____	_____
Mass of $BaSO_4$	_____	_____
Formula weight of $BaSO_4$	_____	_____
Moles of $BaSO_4$	_____	_____
Moles of sulfate ion present in unknown Show calculation	_____	_____
Mass of sulfate ion present in unknown Show calculation	_____	_____
Percent by mass of sulfate ion in unknown Show calculation	_____	_____

POSTLABORATORY QUESTIONS AND CONCLUSIONS

1. What would be the effect on your percentage sulfate determination if:

    a. too little barium chloride solution was used? Explain.

    b. the sample was not thoroughly dried? Explain.

    c. the tube and contents were not cool before the final weighing? Explain.

2. Discuss experimental errors as they apply to your experiment and identify those portions of the procedure where extra care is needed to ensure satisfactory results.

3. The purpose of adding hydrochloric acid in Part 2 is to remove any carbonate ions that might be present in the unknown, so that barium carbonate will not precipitate along with the sulfate. Write the net-ionic equation for the reaction between protons (hydrogen ions) in solution and dissolved carbonate ions.

4. What effect on your sulfate percentage could result if the addition of acid were omitted? Explain.

5. If your unknown is a hydrate, determine the number of waters of hydration and write the correct molecular formula for the hydrated salt.

6. A solution of an alkali metal sulfate contains 0.071 g of the unknown. On reaction with barium chloride there is formed 0.117 g of barium sulfate. What is the identity of the unknown?

EXPERIMENT 12

Carbonate or Bicarbonate?

Introduction

In this experiment you will determine whether an unknown solid is sodium carbonate, Na_2CO_3, or sodium bicarbonate, $NaHCO_3$. This will be done by having the solids react with hydrochloric acid according to either

$$Na_2CO_3(aq) + 2\ HCl(aq) \rightarrow CO_2(g) + 2\ NaCl(aq) + H_2O(l) \qquad \text{(Eq. 12.1)}$$

or

$$NaHCO_3(aq) + HCl(aq) \rightarrow CO_2(g) + NaCl(aq) + H_2O(l) \qquad \text{(Eq. 12.2)}$$

The reaction of an acid with a carbonate is a common reaction. It occurs when an antacid tablet is dissolved in water, or when a baker in bygone days added vinegar to "baking soda" when making a cake.

The law of conservation of matter tells us that the masses of reactants on one side of an equation must equal the masses of the products on the other side of the equation. For that reason, the number of moles of the individual elements (Na, C, O, H, Cl) on one side of an equation must equal those on the other side of the equation.

By similar reasoning the number of moles of sodium carbonate, Na_2CO_3, must equal half the number of moles of sodium chloride according to the first equation. And from the second equation we can see that the number of moles of sodium bicarbonate, $NaHCO_3$, is equal to the number of moles of sodium chloride produced.

A known mass of the unknown will react with an excess of hydrochloric acid and the mass of the sodium chloride produced will be determined. If the unknown were sodium carbonate, then for each mole of the carbonate two moles of sodium chloride would be produced, but if the unknown were sodium bicarbonate, then for each mole of the bicarbonate, one mole of sodium chloride would be produced [see Eqs. 12.1 and 12.2].

The success of this experiment depends on several things. First of all let's look more carefully at the statement that "the unknown will react with an excess of hydrochloric acid." According to the first equation, if our unknown is one mole of sodium carbonate it will require two moles of hydrochloric acid to make the reaction go to completion. If we use less than two moles unreacted sodium carbonate will be left over. If we use, say, three moles of hydrochloric acid, then at the end of the reaction we will have produced one mole of carbon dioxide gas, two moles of sodium chloride, one mole of water, *and* we will have left over one mole of hydrochloric acid.

Hydrochloric acid is a solution of a gas, hydrogen chloride (HCl), dissolved in water. Concentrated hydrochloric acid contains 38% by weight of HCl in water. When this solution is heated to boiling the hydrochloric acid escapes as a gas along with the water. If this is done long enough all of the solution will evaporate.

Carbon dioxide is, as you know, a gas, which in the present experiment can be seen forming bubbles during the reaction. If our unknown reacts with excess hydrochloric acid and the final reaction mixture is heated to over 100°C all of the carbon dioxide will leave, along with the water and the excess hydrochloric acid. The sodium chloride will not evaporate because its boiling point is 1413°C!

When you look up the properties, including the formula weights, of unknowns in a reference book such as the *The CRC Handbook of Chemistry and Physics* you will find that sodium carbonate can take several forms, depending on how many molecules of water are crystallized with it. These substances containing *water of crystallization* are called *hydrates*. You will be supplied with material that has no water of hydration. It is said to be *anhydrous*. Sodium bicarbonate does not form such hydrates.

The objectives of this experiment are to observe the reaction between a typical acid, hydrochloric acid, and either a carbonate or bicarbonate, then to use physical data from the reaction to decide whether the unknown was sodium carbonate or sodium bicarbonate. The mass, and thus the moles, of sodium chloride can be found by simply weighing the tube empty, and then again after the reaction is complete and all volatile residues have been evaporated. We also wish to relate the number of moles of sodium chloride to the number of moles of the unknown assuming that it is either sodium carbonate or sodium bicarbonate, and finally to determine whether the unknown substance is sodium carbonate or sodium bicarbonate.

It is a good idea to answer the four Prelaboratory Questions before carrying out the experiment. In this way you will know the approximate volume of hydrochloric acid needed to react with the two possible compounds.

Safety Information

1. **Safety goggles must be worn at all times in the laboratory.**
2. **Wear a laboratory apron.**
3. **Handle hydrochloric acid with care.** It is very corrosive. Wipe up spills, even of a single drop, immediately with a damp sponge.

Procedure Summary

The objective of this experiment is to determine whether an unknown is either sodium carbonate or sodium bicarbonate. A known amount of unknown is reacted with excess hydrochloric acid. From the weight of the sodium chloride formed it is possible to determine whether the unknown was a carbonate or bicarbonate.

Prelaboratory Assignment

Read the Introduction and Procedure sections carefully and answer the Prelaboratory Questions on the Report Sheet.

Experiment 12   Carbonate or Bicarbonate?   95

Materials

Apparatus

Reaction tube
Hot sand bath
Plastic funnel

Reagents

Unknown (sodium carbonate or sodium bicarbonate)
Boiling chip
Concentrated hydrochloric acid (12 M) to be dispensed with a Beral pipette
Filter paper

Procedure

1. Using the top-loading balance, determine the mass of a clean, dry reaction tube that contains a boiling chip. Using a plastic funnel to aid the transfer, measure into the reaction tube a very small amount of the unknown. A sample near the balance will show you a tube containing the correct amount. Record the mass of the reaction tube plus the unknown and boiling chip.

2. Add 1 drop of concentrated hydrochloric acid to the solid. Note what happens. Add another drop of acid and agitate the tube to mix the contents. Continue adding the acid one drop at a time until no more reaction occurs. To guarantee that all of the unknown has reacted, add one or two additional drops of acid ("an excess"). At this time the aqueous solution contains sodium ions, chloride ions, hydrogen ions from the excess hydrochloric acid, and a little dissolved carbon dioxide.

3. Heat the tube gently on the sand bath or on a hot plate to drive away the water and any excess hydrogen chloride. Avoid spattering the sodium chloride. If necessary, insert a rolled piece of filter paper into the top of the reaction tube in order to absorb the last bit of water condensed in the top of the tube. The tube should be perfectly dry at the end of the experiment. Allow the tube to cool to room temperature and then determine its mass.

Cleaning Up

Because the residue is just sodium chloride and a boiling chip, the sodium chloride can be dissolved from the tube with water and flushed down the drain. Put the boiling chip in the waste basket.

Name _____   Section _____

Lab Instructor _____   Date _____

EXPERIMENT 12   Carbonate or Bicarbonate?

PRELABORATORY QUESTIONS

1. How many moles of sodium carbonate would be in 0.200 g of sodium carbonate?

2. Calculate the molarity (moles per liter) of concentrated hydrochloric acid. As noted in the Introduction, it contains 38% by weight of HCl dissolved in water. The density of concentrated hydrochloric acid is 1.200 g/mL.

3. How many milliliters of concentrated hydrochloric acid are necessary to just neutralize 0.200 g of sodium carbonate?

4. How many milliliters of concentrated hydrochloric acid are necessary to just neutralize 0.200 g of sodium bicarbonate?

DATA AND CALCULATIONS

Mass of reaction tube plus boiling chip   _____

Mass of reaction tube, boiling chip, and unknown   _____

Mass of unknown   _____

Experiment 12   Report Sheet

Moles of unknown assuming it is sodium carbonate   _____

    The molecular weight of anhydrous sodium carbonate is 106 g/mol. Show calculation.

Moles of unknown assuming it is sodium bicarbonate   _____

    The molecular weight of anhydrous sodium bicarbonate is 84 g/mol. Show calculation.

Mass of reaction tube, boiling chip, and sodium chloride _____

Mass of reaction tube plus boiling chip   _____

Mass of sodium chloride   _____

Moles of sodium chloride   _____

Moles of sodium ion   _____

Weight of unknown if it were sodium carbonate   _____

Weight of unknown if it were sodium bicarbonate   _____

Identity of unknown   _____

CONCLUSION

1. Describe your reasoning in arriving at the identity of the unknown.

POSTLABORATORY QUESTION

1. If the unknown was not anhydrous sodium carbonate, but contained some water, would it still be possible to distinguish it from sodium bicarbonate? Show your calculations.

EXPERIMENT 13

Hydrates: Structure and Properties

Introduction

There is a good chance that you are sitting near several hundred pounds of hydrates at this very moment, because both plaster and cement, two of the most important building materials, are hydrates and obviously of great commercial importance.

The object of this experiment is to study the formation and decomposition of hydrates, chemical substances of definite composition formed from, most commonly, ionic compounds and integer or half-integer numbers of water molecules. A common hydrate is bright blue copper sulfate pentahydrate, $CuSO_4 \cdot 5\ H_2O$. The dot indicates that water molecules are bound to the copper and sulfate ions in a definite proportion. Water bound in this manner is called a *ligand*.

Plaster is made by adding water to plaster of Paris, anhydrous calcium sulfate, $CaSO_4$, to form the dihydrate, $CaSO_4 \cdot 2\ H_2O$ in an exothermic reaction. The plaster "sets" rapidly to a highly interlocked crystalline form that has water molecules arrayed between layers of calcium and sulfate ions. Used since antiquity, plaster walls are still standing in Pompeii, and before the invention of cement, plaster was the mortar used between bricks.

Plaster of Paris is made from the mineral gypsum, a hard dense form of the dihydrate, by dehydration (at about 200°C). Another mineral form of calcium sulfate dihydrate is alabaster, used for many centuries as a sculpture medium.

Anhydrous calcium chloride, $CaCl_2$, avidly picks up water to eventually form a hexahydrate (six waters of hydration). The reaction is exothermic and the resulting solution has a very low freezing point, so the primary use of this salt is for deicing roads, which it can do at temperatures down to –51°C. It is equally useful in the summer when it will remove moisture from the air to form a damp solid. As such it is used to control dust in highway construction. Anhydrous calcium chloride is commonly used in the laboratory as a desiccant where it will remove moisture from air (in a desiccator) or from gases that flow through particles of the solid. It also has the property of removing water from organic liquids such as diethyl ether that can dissolve about 8% of its weight of water.

Copper sulfate pentahydrate is blue, whereas the anhydrous salt is white. Cobalt chloride in moist air is present as the hexahydrate, $CoCl_2 \cdot 6\ H_2O$, which is red, while in dry, cool air it is the blue anhydrous salt, and at intermediate humidities is present as the purple dihydrate, $CoCl_2 \cdot 2\ H_2O$. As such it serves as a humidity indicator. For example, it is incorporated, with the desiccant silica gel; when it turns pink the desiccant needs to be reactivated by heating. Some hydrates lose water at room temperature and normal humidity; they are said to *effloresce*. Other salts, such as calcium chloride, pick up so much moisture from the air that they dissolve in the resulting liquid. These salts are *deliquescent*.

A very large class of organic compounds is called carbohydrates. These are not real hydrates of carbon. This misnomer arose in the middle of the last century when only the empirical formulas, $[C(H_2O)]$, of many sugars were known. Glucose, for example, which has a molecular formula of $C_6H_{12}O_6$, can be written as $C_6(H_2O)_6$, but this sugar is not carbon hexahydrate. When a sugar is heated it will lose the elements of water, but in the process will decompose irreversibly.

100   Experiment 13   Hydrates: Structure and Properties

True hydrate formation is completely reversible. The partial dehydration of sucrose, $C_{12}H_{22}O_{11}$, leads to taffy, caramel, and finally to the dark brown substance from which peanut brittle is made.

Procedure Summary

You will investigate the nature of hydrated and anhydrous salts and observe the phenomena of efflorescence and deliquescence. Then you will determine the number of molecules of water of hydration in an unknown and investigate the nature of some unknown hydrates and a carbohydrate.

Prelaboratory Assignment

Read the Introduction and Procedure sections carefully, and answer the Prelaboratory Question on the Report Sheet.

Materials

Apparatus

Milligram balance
Electrically heated sand bath
10×75-mm culture tubes
Glass stirring rod
1-mL Beral pipette
Weighing paper cut into small squares
6-in. watchglass
Wire test tube clamp

Reagents

Anhydrous cobalt chloride, $CoCl_2$
Anhydrous calcium chloride, $CaCl_2$
Sodium carbonate decahydrate, $Na_2CO_3 \cdot 10\ H_2O$
Potassium aluminum sulfate, (alum), $KAl(SO_4)_2 \cdot H_2O$
Anhydrous copper sulfate, $CuSO_4$
Calcium sulfate
Copper sulfate pentahydrate, $CuSO_4 \cdot 5\ H_2O$
Sucrose (table sugar)
Cobalt chloride
Potassium chloride
Nickel chloride
Potassium carbonate
Unknown hydrate

Experiment 13   Hydrates: Structure and Properties   101

> **Safety Information**
> 1. **Safety glasses must be worn at all times in the laboratory.**
> 2. **The sand bath is very hot.** Avoid touching it.

Procedure

1. Turn on the electrically heated sand bath to a setting of 50% to 60% (5 or 6). This will produce temperatures exceeding 250°C in the sand.

Part 1

2. On five individual small (1.5 × 1.5-cm) pieces of paper carefully weigh about 300 mg of each of the following substances. Label the papers with the formulas of the substances:

   Anhydrous cobalt chloride, $CoCl_2$
   Anhydrous calcium chloride, $CaCl_2$
   Sodium carbonate decahydrate, $Na_2CO_3 \cdot 10\ H_2O$
   Potassium aluminum sulfate, alum, $KAl(SO_4)_2 \cdot H_2O$
   Anhydrous copper sulfate, $CuSO_4$

3. Place the papers very close to each other and cover them with a watchglass. Note the appearance of each salt at the beginning and at the end of the laboratory period. Just before the end of the period weigh each salt again to determine if it has lost or gained weight or remained unchanged.

Part 2

4. Place a small amount of each of the following substances into seven separate 10 × 75-mm culture tubes (use only enough to cover the bottom of the tube). Gently crush the solids to a fine powder with a glass stirring rod if necessary.

   Calcium sulfate
   Copper sulfate
   Sucrose (table sugar)
   Cobalt chloride
   Potassium chloride
   Nickel chloride
   Potassium carbonate

5. Heat the bottom of each tube in a hot sand bath and observe any changes. If the compound loses water on heating, moisture should collect on the upper portion of the tube, which should be cool enough to hold in your fingers. Alternatively, the tube can be held in a wire test tube holder. This is an indication the compound may be a hydrate. Note the appearance of the solid after heating.

6. After the tubes cool add about a milliliter of water to each one using a graduated Beral pipette, stir, and observe what happens. If the substance is a hydrate it should dissolve and give a solution close in color to the original hydrated salt.

Part 3

7. Mark, perhaps with scratches, three 10×75-mm culture tubes. Carefully weigh (± 0.001 g) about 0.4 g of an unknown hydrate into each one. Gently crush the material to a fine powder with a glass rod. Push these tubes well down into a hot sand bath and leave them there for about 15 minutes or longer. Be sure any drops of water that may collect on the upper parts of the tubes initially are all gone when they are removed from the heat.

8. Use a test tube holder or tongs to remove the tubes from the sand bath. Allow the tubes to cool to room temperature, carefully wipe off any adhering grains of sand, and weigh the tubes carefully. The unknown you have been given, when anhydrous, has a molecular weight of 136.14 g/mol. You are to determine how many waters of hydration the hydrate contains.

Cleaning Up

All of the substances used in this experiment except cobalt chloride are relatively harmless and can be flushed down the drain with water. All cobalt chloride samples should be placed in the receptacle provided for recycling.

Name _____ Section _____

Lab Instructor _____ Date _____

EXPERIMENT 13   Hydrates: Structure and Properties

PRELABORATORY QUESTION

1. If an unknown hydrate were not heated sufficiently to drive off all of the water of hydration, what would be the effect on the calculation of the number of waters of hydration? Explain.

DATA AND OBSERVATIONS

Part 1

	Initial Weight	Final Weight
Anhydrous cobalt chloride, $CoCl_2$	_____	_____
Appearance: initial and final _____		
Anhydrous calcium chloride, $CaCl_2$	_____	_____
Appearance: initial and final _____		
Sodium carbonate decahydrate, $Na_2CO_3 \cdot 10\ H_2O$	_____	_____
Appearance: initial and final _____		
Potassium aluminum sulfate, alum, $KAl(SO_4)_2 \cdot H_2O$	_____	_____
Appearance: initial and final _____		
Anhydrous copper sulfate, $CuSO_4$	_____	_____
Appearance: initial and final _____		

Part 2

	Initial Color	Water Evolved on Heating?	Final Color	Soluble in Water?	Is It a Hydrate?
Calcium sulfate	_____	_____	_____	_____	_____
Copper sulfate	_____	_____	_____	_____	_____
Sucrose	_____	_____	_____	_____	_____
Cobalt chloride	_____	_____	_____	_____	_____
Potassium chloride	_____	_____	_____	_____	_____
Nickel chloride	_____	_____	_____	_____	_____
Potassium carbonate	_____	_____	_____	_____	_____

Part 3

	Trial 1	Trial 2	Trial 3
Weight of tube plus salt (about 0.4 g)	_____	_____	_____
Weight of empty tube	_____	_____	_____
Weight of salt before heating	_____	_____	_____
Weight of salt after heating	_____	_____	_____
Number of moles of anhydrous salt (formula weight 136 g/mol)	_____	_____	_____
Weight of water evolved during heating	_____	_____	_____
Number of moles of water evolved (formula weight 18.02 g/mol)	_____	_____	_____
Ratio of moles of salt to moles of water	_____	_____	_____
Formula of hydrate, $X \cdot Y\ H_2O$	$X \cdot$ ___ H_2O		

EXPERIMENT 14

Periodic Trends in Physical Properties: Densities of Unknowns

Introduction

One of the triumphs of Mendeleev's design of the periodic table was its prediction of the existence and properties of elements that had not yet been discovered. Gallium, atomic number 31, and germanium, atomic number 32, are good examples. Aluminum, silicon, and tin were known, as were zinc and arsenic. But Mendeleev recognized that there should be elements with chemical properties similar to the properties of silicon and aluminum, but with atomic masses between those of zinc and arsenic. He had such faith in their existence that he even named these missing elements, calling them *eka-aluminum* and *eka-silicon* (from the Sanskrit word *eka*, "first") because they should appear first after aluminum and silicon in the periodic table. In this experiment you will attempt to predict one of the properties of Mendeleev's eka-silicon by measuring the densities of the elements above and below it, then taking an average of the two values.

In the second part of the experiment you will measure the density of some U.S. one-cent pieces and compare results with the expected values. The volume of the pennies will be measured by water displacement as well as calculated from their dimensions.

Procedure Summary

The density of irregularly shaped solids is found by water displacement. Water (or other liquid) is placed in a graduated cylinder and its volume is precisely measured, then a known mass of the solid is added. The volume of the solid is found by subtracting the final liquid level from the initial reading, and density is calculated according to the following equation which says that density is equal to the ratio of mass to volume:

Density = mass/volume

Prelaboratory Assignment

Read the Introduction and Procedure sections, and answer the Prelaboratory Question on the Report Sheet.

Materials

Apparatus

10-mL graduated cylinder
Top-loading balance

Reagents

Silicon, granular or small pieces
Tin, shot or granular
Pennies

Safety Information

1. **Safety goggles must be worn at all times in the laboratory.**

Procedure

Part 1

Use the same 10-mL graduated cylinder for both of your density determinations. Each solid element's density is found in the manner described below.

1. Fill the graduated cylinder roughly halfway with water and record the volume to the nearest 0.1 mL (Figures 14.1 and 14.2). Determine the mass of your container of solid pieces to the full precision of your balance, then add enough of the solid to raise the liquid level by at least 3.0 mL but not above the 10.0-mL line (why not?). Be sure to add the solid carefully, to minimize splashing. Any drops of water that cling to the sides of the graduate will diminish the accuracy of your volume readings. The container with unused solid is reweighed and the mass of solid used is found by subtraction. Record the final volume of solid and water.

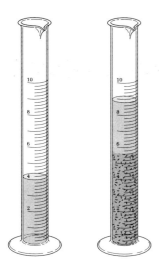

Figure 14.1 The volume of the irregularly shaped pieces in the cylinder on the right is given by the difference between the initial and final readings of the liquid level.

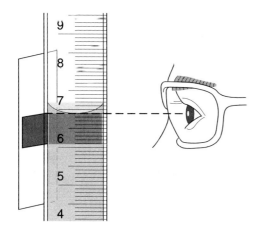

Figure 14.2 Avoid parallax error. Line of sight should be horizontal. A dark area on a white card will cast a dark reflection on the bottom of the meniscus, making it easier to see.

2. Repeat the process with the other solid. The graduated cylinder should be rinsed clean after each trial, as described in the Cleaning Up section.

Part 2

3. Using the centimeter ruler printed in the Appendix, measure the diameter of a penny. Also measure the height of 25 pennies and then calculate the volume of the resulting cylinder of pennies.

4. Fill a 25-mL graduated cylinder about one-third full of water, measure the volume of the water to the nearest 0.1 mL, and then add to the cylinder the 25 U.S. one-cent pieces and measure the volume again, making sure no air bubbles are trapped between the pennies.

Cleaning Up

Do not empty your graduated cylinder into the sink. The pieces of solid are collected separately in strainers; they will be dried and used again. Dump the entire contents of your graduated cylinder into the appropriate strainer, using extra water if necessary to get all of the solid out.

Table 14.1 **Densities of Some Metals**

	g/cm^3
Aluminum, Al	2.70
Copper, Cu	8.96
Germanium, Ge	5.32
Gold, Au	19.3
Iron, Fe	7.87
Magnesium, Mg	1.74
Nickel, Ni	8.90
Silver, Ag	10.5
Zinc, Zn	7.13

Name _____ Section _____

Lab Instructor _____ Date _____

EXPERIMENT 14 Periodic Trends in Physical Properties: Densities of Unknowns

PRELABORATORY QUESTION

1. The density of calcium is 1.54 g/cm^3 and the density of barium is 3.51 g/cm^3. Calculate the expected density of strontium. The reported density of strontium is 2.6 g/cm^3. What is the percent error in the calculated density?

DATA

Part 1

Initial volume of water in cylinder	_____ mL
Initial mass of container of solid silicon	_____ g
Volume of water after adding silicon	_____ mL
Final mass of silicon container	_____ g
Initial volume of water in cylinder	_____ mL
Initial mass of container of solid tin	_____ g
Volume of water after adding tin	_____ mL
Final mass of tin container	_____ g

Part 2

Diameter of one penny	_____ cm
Height of 25 pennies	_____ cm
Mass of pennies	_____ g
Final level of water in cylinder	_____ mL

109

Initial level of water in cylinder _____ mL

Volume of pennies in cylinder _____ mL

CALCULATIONS AND CONCLUSIONS

1. Calculate the densities of silicon and tin.

2. The density of germanium (eka-silicon) is assumed to be about midway between the densities of silicon and tin. Calculate the value of germanium's density as predicted by your results for the other two elements.

3. Calculate the percent error between your theoretical and the actual density of germanium. Look up the density of germanium in *The CRC Handbook of Chemistry and Physics*. To calculate your percent error, determine the absolute value of the difference between your answer to Question 2 and the theoretical value, then divide that difference by the theoretical value. This will give you a decimal fraction, which you can convert to a percent by multiplying the fraction by 100.

4. What is the volume of pennies calculated from $V = \pi r^2 h$ where r is the radius and h is the height of the cylinder of pennies, as measured in Step 3?

   a. Volume of cylinder of pennies (calculated) _____

   Show calculations

   b. Volume of cylinder of pennies as measured in graduated cylinder _____

Experiment 14   Report Sheet   111

    c.   Difference between calculated and measured volumes of pennies _____

    d.   Density of pennies calculated from mass of pennies and
          volume of pennies measured in graduate (4b) _____

    e.   Density of pennies calculated from mass of pennies and
          volume of penny cylinder from 4a. _____

5.   a.   In which value for the density of pennies (4d or 4e) do you have greater confidence? Explain.

    b.   Determine the percentage deviation between the two density values by dividing the difference between them by the value you selected in 5a. Convert your result to a percentage.

6.   How does your calculated value for the density of pennies (Question 4d and 4e) compare with the densities of metals reported in Table 14.1? Discuss your results.

Experiment 14   Report Sheet

POSTLABORATORY QUESTIONS

1. Would you expect to get the same results using a different liquid in place of water? Discuss the factors that would influence your choice of liquid to be used.

2. You have a density of about 1.0 g/mL. Sea water has a density greater than one, while fresh water has a density of 1.0 g/mL. All other things being equal would you expect to float or sink in sea water compared to fresh water?

3. A piece of ice, density $0.9 g/cm^3$, weighing 3.0g is placed in a graduated cylinder half full of ice water. Before any of the ice melts how far, in mL, will the water level rise in the cylinder? After the ice melts how far will the water level rise in the cylinder?

EXPERIMENT 15

Distillation

Introduction

When a liquid is heated the vapor pressure of the liquid, the tendency of molecules to escape from the surface, increases until it becomes equal to the atmospheric pressure, at which point the liquid begins to boil. When this process is carried out in a distilling apparatus (Figure 15.1) the vapor from the boiling liquid rises from the flask, warms the distillation head and the thermometer and flows down the condenser, where it loses heat to the surroundings and condenses back to a liquid. This process of distillation is used for purification and separation of substances. A pure liquid will have a constant boiling point as long as the pressure over the liquid is constant. If a packing, such as a copper sponge, is placed in the distillation column, much better separation results.

If the flask held seawater, for instance, the distillate, which is the water vapor that condenses and is collected in the receiver, would be pure water and the salt, which is not volatile, would be left behind.

If the flask contained two miscible liquids such as methyl alcohol, CH_3OH, and water, H_2O, the distillate would consist of the lower boiling liquid, methyl alcohol [boiling point, (bp) 64.6°C], which distills first, followed by the higher boiling liquid, water (bp 100°C).

It is assumed that the distillation in this experiment will take place at atmospheric pressure, 101 kPa or 760 mm Hg. If the distillation were to take place in Mexico City at an elevation of 7,700 ft (2,310 m) the boiling point of water would be only 92.8°C because the bp decreases as the pressure above the liquid decreases.

Very often pure, clean liquids can become heated above their boiling points, a process called *superheating*. Then they may suddenly boil, sometimes with almost explosive violence. To prevent this from happening, a boiling chip or boiling stick is added to the solution. The chip and the wood stick have tiny bubbles of air trapped in them which when heated are released and form the nucleus on which bubbles of vapor form, allowing smooth boiling of the liquid at its boiling point. A new boiling chip must be added every time a liquid is allowed to cool and again brought up to the boiling point.

Procedure Summary

In the first part of this experiment methyl alcohol will be separated from water that contains a colored, nonvolatile impurity by the process of distillation. Without going into the theory of distillation, this experiment is presented as an example of an important method of purification of substances.

In the second part of this experiment you will separate by fractional distillation two unknown liquids that have boiling points at least 20°C apart. The object of the experiment is to determine the boiling points of the two liquids and thus identify them.

Boiling point will be recorded and plotted as a function of drops of distillate, which can be equated to volume of distillate. Both of these experiments can be interfaced to a computer.

Prelaboratory Assignment

Read the Introduction and Procedure sections carefully, and answer the Prelaboratory Question on the Report Sheet.

Materials

Apparatus

Sand bath
Distilling apparatus
    5-mL long-necked round-bottomed flask
    Connector with support rod
    Distillation head/condenser
    Thermometer adapter
1-mL graduated pipette
Thermometer, mercury, digital or digital probe
Vials, 2
30-mL beaker
Three-prong clamps
Clamp holders
Ring stand
Metal spatula
Metal sponge, 1 g

Reagents

Part 1:
Methyl alcohol, 2 mL
Water, 2 mL
Boiling chip

Part 2:
A mixture taken from the compounds in Table 15.1.

Table 15.1 Boiling Points for Various Liquids

	bp, (°C)
Acetone, CH_3COCH_3	56.5
2-Propanol, $CH_3CHOHCH_3$	82.4
Water, H_2O	100
1-Butanol, $HOCH_2CH_2CH_2CH_3$	117.7

Safety Information

1. **Safety goggles must be worn at all times in the laboratory.**
2. **Keep flames away from the distilling apparatus.**
3. **The sand bath can get very hot so avoid touching it.**
4. **Use great care if you are using a mercury thermometer.** Prevent it from rolling onto the floor or becoming broken. Never place the mercury or digital thermometer in the sand bath. Should the mercury thermometer become broken notify your teacher at once. Mercury is very poisonous.
5. **Methanol (methyl alcohol) and 1-butanol (*n*-butyl alcohol) are toxic.**

Procedure

Part 1: Distillation of Water/Methanol Mixture.

1. Turn on the electricity to your sand bath if it has not already been done. If you are using a Thermowell as your heat source set the controller to 3. Place a boiling chip in the one-piece distilling apparatus. Add 2.0 mL of methyl alcohol and then 2.0 mL of water to the flask using a 1-mL graduated pipette. Push a 1-g piece of copper sponge down until the

top surface of it is about 2.5 cm below the sidearm and just above the flask (Figure 15.1). Attach the thermometer adapter to the distilling head/condenser and then carefully push the glass thermometer down through the thermometer adapter. Hold the thermometer near the insertion point and lubricate it with a drop of water to make this process easier. Be sure the bulb or tip of either the mercury or digital thermometer is *completely below* the sidearm of the condenser as seen in Figure 15.1; otherwise the mercury in the bulb or the thermistor will not be heated completely by the distilling vapors.

Both the digital thermometer and the digital sensor for the computer will have bushings on them to give a tight fit in the thermometer adapter.

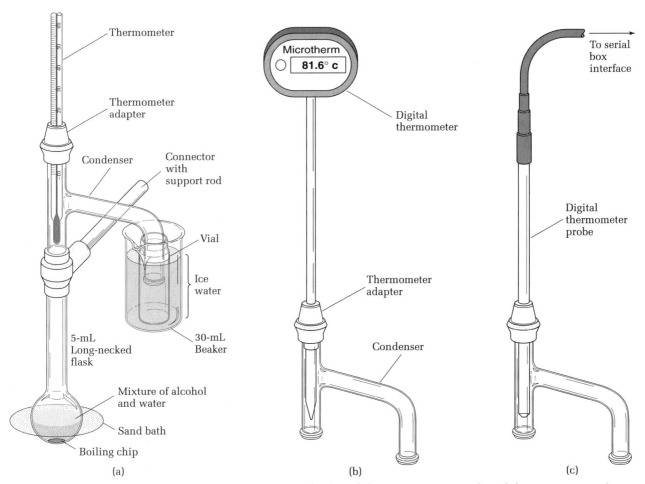

Figure 15.1 Distillation apparatus, with (a) mercury, (b) digital thermometer, or (c) digital thermometer probe.

2. Clamp the whole apparatus to a ring stand so that the flask is just barely in the sand. Clamp a 30-mL beaker containing a vial in ice water as seen in Figure 15.1. Gently support the thermometer with a clamp so that the entire apparatus is vertical and the bulb of the thermometer does not touch the distilling adapter.

If boiling does not begin immediately, pile sand around the bottom of the flask using your metal spatula. If the boiling is too vigorous scrape sand away from the bottom of the flask. Change receiving vials when the temperature jumps to about 80°C.

3. Note volumes and boiling point of first and second fractions and the volume of the pot residue. Do not distill to dryness. The container from which a distillation takes place is called the *pot* and the container into which the distillate drips is called the *receiver*.

   In the hood, as an optional experiment, try to set fire to a mixture of 50% methyl alcohol and 50% water as well as your first fraction. Place samples in an evaporating dish for this experiment. A mixture of 50% ethanol and 50% water will just barely burn when set afire. This was an early test for the purity of distilled spirits. If it burned it was proof that it contained half alcohol, or was "100 proof."

Cleaning Up

Empty the first distillate in the container provided for recovered methanol. Take off the thermometer and adapter from the apparatus and with a hook made of a piece of wire, pull out the copper sponge. Do not discard it. Pour the pot residue (water) down the drain, being careful to recover and dispose of the boiling chip. Shake out or otherwise remove as much of the water from the apparatus as possible and then rinse the entire apparatus with a few drops of acetone to remove the last traces of water. Blot the copper sponge as dry as possible and then rinse it with a few drops of acetone, blot again, and allow to dry. The apparatus and sponge must be dry before proceeding with the next part of the experiment.

Part 2: Distillation of Two Unknown Liquids

4. Add to the 5-mL long-necked round-bottomed flask 4 mL of one of the five unknown mixtures, A, B, C, D, or E, and a boiling chip. Assemble the apparatus exactly as before. After adding the unknown and the boiling chip the long neck should be stuffed with 1 g of copper sponge. Carry out the distillation in the same way as before, except do not change the vials halfway through the distillation. Simply record drop number versus observed boiling point. Remember that the slower the distillation is carried out, the better the separation. Do not distill to dryness.

Computer Interface

If you are interfacing this experiment with a computer and are using the Vernier apparatus,[1] the temperature probe and the 9-V power supply should be connected to the Vernier Serial Box Interface, which in turn is connected to a serial port, usually the modem port, of the computer. Load and start the computer program called Data Logger and open the Direct Connect Temperature Probe file in the Experimental Files folder. Under the Data menu select New Column and under the pull-down menu titled Formula select Prompted and hit OK. You will then select the Event Mode as directed. On the main menu under Display select Axes and on the resulting display select Temperature 1 and set the temperature range from 55 to 105°C. At the top right there should be an X only in the Data A box. Do not change the Column 3

---

[1] Vernier, 8565 S.W. Beaverton-Hillsdale Hwy., Portland, OR 97225-2429
http://www.teleport.com/~vernier, email dVernia@vernia.com

information. It is set for 100 events (drops). Under Collect on the main menu you should find Event Mode and Select Display Inputs checked. If they are not, select them.

Start the distillation. As soon as the thermocouple temperature, displayed at the bottom of the graph, reads above 55°C select Start at the bottom left corner of the graph. As the first drop falls from the sidearm of the distilling head select Keep with the mouse button, type 1 under Column 3, and select OK. This records the first temperature on your graph. As the second drop falls select Keep again and then enter 2 for the second drop temperature and hit OK. Repeat this sequence for each drop. You can copy your resulting graph to another document, save it, print it, or save only the data table from the 60 or so drops that you collected.

Cleaning Up

Place the pot residue, boiling chip, and the distillate in the appropriate collection container. Allow the apparatus to drain dry. It should not be necessary to wash it.

Name _____ Section _____

Lab Instructor _____ Date _____

EXPERIMENT 15   Distillation

PRELABORATORY QUESTIONS

1. From your experience and a knowledge of what distillation entails, name something besides liquor and gasoline that has been prepared or purified by distillation.

2. Why is it necessary that the entire apparatus be dry before you begin step 4? Why was acetone chosen for rinsing rather than water?

DATA AND OBSERVATIONS

Part 1

Volume of water added to flask _____

Volume of methanol added to flask _____

*If a computer–interfaced probe is used, affix a printout of your data to this report.

First Fraction Drop Number	Observed bp	First Fraction Drop Number	Observed bp
_____	_____	_____	_____
_____	_____	_____	_____
_____	_____	_____	_____
_____	_____	_____	_____
_____	_____	_____	_____
_____	_____	_____	_____

Experiment 15   Report Sheet

First Fraction Drop Number	Observed bp	First Fraction Drop Number	Observed bp
_____	_____	_____	_____
_____	_____	_____	_____
_____	_____	_____	_____
_____	_____	_____	_____
_____	_____	_____	_____
_____	_____	_____	_____
_____	_____	_____	_____
_____	_____	_____	_____
_____	_____	_____	_____
_____	_____	_____	_____
_____	_____	_____	_____
_____	_____	_____	_____
_____	_____	_____	_____
_____	_____	_____	_____
_____	_____	_____	_____
_____	_____	_____	_____
_____	_____	_____	_____
_____	_____	_____	_____
_____	_____	_____	_____
_____	_____	_____	_____
_____	_____	_____	_____
_____	_____	_____	_____
_____	_____	_____	_____
_____	_____	_____	_____
_____	_____	_____	_____
_____	_____	_____	_____

Experiment 15   Report Sheet   121

Approximate volume of first fraction   _____

Boiling point of major part of first fraction _____

Boiling point of major part of second fraction _____

Color of pot residue   _____

Color of distillate   _____

Probable identity of the liquid left in the distilling flask (the pot residue)

Optional: Result when attempt was made to set first fraction afire in the hood   _____

Optional: Result when attempt was made to set 50:50 methanol/water mixture afire in the hood   _____

Conclusion with regard to ability of distillation to separate methanol from water

Part 2

Initial volume of two-component mixture _____

First Fraction Drop Number	Observed bp	First Fraction Drop Number	Observed bp
_____	_____	_____	_____
_____	_____	_____	_____
_____	_____	_____	_____
_____	_____	_____	_____
_____	_____	_____	_____
_____	_____	_____	_____
_____	_____	_____	_____
_____	_____	_____	_____
_____	_____	_____	_____
_____	_____	_____	_____
_____	_____	_____	_____
_____	_____	_____	_____
_____	_____	_____	_____
_____	_____	_____	_____
_____	_____	_____	_____
_____	_____	_____	_____
_____	_____	_____	_____
_____	_____	_____	_____
_____	_____	_____	_____

Experiment 15 Report Sheet 123

Second Fraction Drop Number	Observed bp	Second Fraction Drop Number	Observed bp
_____	_____	_____	_____
_____	_____	_____	_____
_____	_____	_____	_____
_____	_____	_____	_____
_____	_____	_____	_____
_____	_____	_____	_____
_____	_____	_____	_____
_____	_____	_____	_____
_____	_____	_____	_____
_____	_____	_____	_____
_____	_____	_____	_____
_____	_____	_____	_____
_____	_____	_____	_____
_____	_____	_____	_____

On your own graph paper, make a graph of boiling point (ordinate) versus drop number (abscissa). If a computer-interfaced probe is used, affix data and graph printouts to this sheet.

Boiling point of major part of first fraction   _____

Boiling point of major part of second fraction   _____

Identities of the components of the mixture   _____

POSTLABORATORY QUESTIONS

1. a. What is the effect on the boiling point of a solution (for example, water) produced by a soluble, non-volatile substance (for example, sodium chloride)?

   b. What is the effect of an insoluble substance such as sand?

   c. What is the temperature of the vapor above these two boiling solutions?

2. In the distillation of a pure substance (for example, water), why does not all of the water vaporize at once when the boiling point is reached?

3. Why is it dangerous to attempt to carry out a distillation in a completely closed apparatus, with no vent to the atmosphere.

4. What effect does the reduction of atmosphere pressure have on the boiling point? Can methanol be separated from water at 350 mm Hg instead of 760 mm Hg?

5. Is it possible to make water boil below room temp?

EXPERIMENT 16

Avogadro's Law

Introduction

Avogadro's hypothesis states the following: *Equal volumes of gas at the same temperature and pressure contain equal numbers of molecules.* The hypothesis may be generalized to say that, at constant temperature and pressure, the volume occupied by a gas is directly proportional to the number of molecules (or moles of molecules) it contains: $V_{P,T} \propto n$. This expression is often referred to as *Avogadro's law.* This experiment will provide you with an opportunity to test the validity of the relationship for yourself.

Baking soda, sodium bicarbonate ($NaHCO_3$), will react with the acetic acid in vinegar to produce carbon dioxide gas, along with an aqueous solution of sodium acetate:

$$NaHCO_3(s) + CH_3COOH(aq) \rightarrow Na^+CH_3COO^-(aq) + H_2O(l) + CO_2(g) \qquad (16.1)$$

Prelaboratory Assignment

Read the Introduction and Procedure sections, and answer the Prelaboratory Questions on the Report Sheet.

Procedure Summary

A known mass of sodium bicarbonate is allowed to react with an excess of acid [vinegar (acetic acid)] to generate carbon dioxide. The volume of carbon dioxide generated is measured. Knowing the moles of sodium bicarbonate allows the calculations of the molar volume of carbon dioxide, i.e. the volume in liters occupied by one mole of CO_2 at standard temperature and pressure. The process is carried out a total of five times, with increasing masses of baking soda.

Materials

Apparatus

Milligram balance
Bottle and syringe setup (see Figure 16.1)

Reagents

Sodium bicarbonate
White vinegar
Glycerol (for lubricating syringe)

Safety Information

1. **Safety goggles must be worn at all times in the laboratory.** While neither the reactants nor products are considered hazardous, normal safety precautions should be practiced.

Procedure

The apparatus we will use, shown in Figure 16.1, consists of a small clear, hard plastic bottle and a 60-mL syringe. The syringe is connected by a short tube to the cap of the bottle. A known mass of sodium bicarbonate is placed in the bottom of the bottle. The vinegar is placed in a 13 × 100-mm culture tube, then carefully lowered into the bottle so that none of the liquid is spilled. The cap is carefully replaced and screwed down snugly. Tilting the bottle allows the vinegar to leave the tube and react with the baking soda, as shown in Eq. (16.1). As the $CO_2(g)$ is formed, it causes the plunger of the syringe to move upward until the pressures inside and outside the syringe are the same. By varying the amount of sodium bicarbonate, you can establish a relationship between moles of carbon dioxide generated and the volume it occupies.

1. Use a piece of clear plastic or rubber tubing (about 5 to 7 cm) to connect the glass tubing on the bottle cap to the 60-mL syringe. Clamp the syringe loosely on a ring stand. Weigh to ±1 mg about 0.15 g of sodium bicarbonate into a small vial, which is placed in the plastic bottle of the apparatus (Figure 16.1). Alternatively, the solid can be weighed directly into the bottle if it has been rinsed well.

Figure 16.1 Apparatus for Avogadro's law experiment.

2. Fill a 100-mm tube three-fourths full with vinegar. Lower the reaction tube carefully into the bottle, being careful not to spill any on the baking soda already in the bottle.

3. Set the syringe to 0 mL, then screw the bottle into the cap attached to the syringe. Tilt the bottle, allowing the vinegar to pour out of the tube. As the reaction proceeds and gas is produced, the syringe's piston will begin to move up. Gently push and release the piston to be sure it does not stick. As the piston slows down, flick the bottle several times with your finger to ensure that all the baking soda has reacted.

4. When the reaction is complete, record the volume of gas produced, unscrew the bottle from the cap, remove and rinse the culture tube and the bottle, and repeat the experiment, varying the amount of baking soda. Do not exceed the mass of $NaHCO_3$ that you calculated in Prelab Question 1. Carry out a minimum of five trials.

Cleaning Up

There are no environmental hazards in this experiment. At the conclusion of each trial, the contents of the bottle may be rinsed down the drain with water.

Name _____   Section _____

Lab Instructor _____   Date _____

EXPERIMENT 16   Avogadro's Law

PRELABORATORY QUESTIONS

1. Calculate the mass of baking soda that would generate 60 mL of carbon dioxide, assuming 1.0-atm pressure and a temperature of 20°C. This is the maximum mass that can be used with a 60-cm^3 syringe.

2. Look up the solubility of carbon dioxide in water at 25°C and at atmospheric pressure in *The CRC Handbook of Chemistry and Physics*. Assuming that such a solution has a density of 1.00 g/mL, calculate the molar solubility of carbon dioxide at 25°C.

3. a. If the 100-mm culture tube holds 10.0 mL of vinegar, calculate the number of moles of carbon dioxide gas that would dissolve in this volume at 25°C. Assume that the solubility of CO_2 is the same in vinegar as it is in water.

   b. What volume would be occupied at 25°C and 1.00 atm by the number of moles you calculated in Question 3a? This is the amount by which the volume change you observe in the experiment will fall below the theoretical value.

Experiment 16   Report Sheet

DATA TABLE

Mass of NaHCO$_3$ (g)	Volume of CO$_2$ (mL)	Mass of NaHCO$_3$ (g)	Volume of CO$_2$ (mL)
___	___	___	___
___	___	___	___
___	___	___	___
___	___	___	___
___	___	___	___

CALCULATIONS

1. For each trial you conducted, use the mass of sodium bicarbonate to calculate the moles of CO$_2$ generated.

2. Convert your results from the preceding question 1 to values for the molar volume of CO$_2$ for each trial, in L/mol. Calculate an average value, including average deviation.

CONCLUSIONS

Refer to your Prelaboratory Questions. What volume of CO$_2$ can be expected to dissolve in the water (vinegar) solution in the bottle? Discuss the effect this solubility would have on your results. Is it sufficient to explain any deviation between your experimental value for the molar volume of CO$_2$ at prevailing conditions and that predicted by the ideal gas law? Explain.

EXPERIMENT

17 Charles' Law: Temperature/Volume Relationships for Gases

Introduction

In this experiment you will investigate the effect of temperature on a constant volume of gas in an attempt to confirm Charles' law and also determine the value of absolute zero.

Jacques Charles found that the ratio of the volume of a gas to its absolute temperature is equal to a constant, k:

$$V/T = k$$

A gas, air, is confined in a thin glass tube, a melting point capillary, above a little plug of cottonseed oil. As the air expands on heating, the plug of oil (Figure 17.1) moves down the capillary; as the air contracts on cooling, the plug moves up. The scale on the thermometer

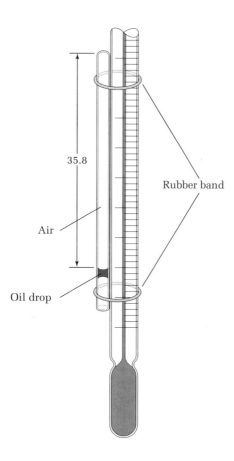

Figure 17.1 Thermometer with capillary attached.

next to the capillary is used as a ruler to measure the length of the gas cylinder in arbitrary units of length. Since the diameter of the capillary is uniform, the length of the air bubble is proportional to its volume. It is assumed throughout this experiment that the pressure on the gas is constant.

When you plot your data on a piece of graph paper you will find that the length of the air column changes in a regular way as the temperature decreases. There is a theoretical point at which the length of the air column is zero. The temperature at this point is absolute zero. Of course, in reality the column can never have zero length (which would correspond to zero gas volume) because long before that point is reached the gas would turn into a liquid and Charles' law would no longer apply. But your data will allow you to calculate a value for absolute zero, the lowest possible temperature. Although it is impossible to ever attain this temperature, scientists have come within a few millionths of a degree of it using special apparatus and techniques.

At absolute zero, motions of atoms nearly cease. Near absolute zero, some of the properties of matter change in interesting and unique ways. Some metals lose their resistance to the flow of electric current. They become superconductors. One form of helium completely loses its viscosity and surface tension. It becomes a superfluid that can flow up and out of a container!

Procedure Summary

The volume of a gas in a very small cylinder is measured as a function of temperature. A plot of the volume versus the absolute temperature on extrapolation will give the value of absolute zero.

Prelaboratory Assignment

Read the Introduction and Procedure sections carefully and answer the Prelaboratory Question on the Report Sheet.

Materials

Apparatus

Thermometer
Sealed melting point capillary
Small rubber bands cut from gum rubber tubing
Beakers with water at five different temperatures: near zero (ice), room temperature, ~20, 40, 60, 80, and close to 100°C

Experiment 17 Charles' Law: Temperature/Volume Relationships for Gases

> **Safety Information**
> 1. **Safety goggles must be worn at all times in the laboratory.**
> 2. **Handle the thermometer with great care.** It is not only expensive, but if it is a mercury thermometer and if it should break the mercury vapor from the thermometer is very poisonous. Should you break the thermometer, notify your teacher at once.
> 3. **Use the beakers of hot water with care.**

Procedure

1. Attach a previously-sealed melting point capillary containing a drop of oil to your thermometer using small rubber bands that have been cut from a piece of gum rubber tubing. Record the temperature on the thermometer and also record the length of the column of air in the capillary, as shown in Figure 17.1. In this case you are using the marks on the thermometer just like a ruler. In the figure the air column is seen to be about 35.8 divisions long.

2. Next, immerse the thermometer with its attached capillary in a beaker of ice water or very cold water. Then put it into four other beakers having water at room temperature and near 20, 40, 60, 80, and 100°C. In each case wait until the temperature reading on the thermometer is fairly constant before recording the length of the column of air.

    If you are using a digital thermometer for this experiment (safer, faster, and more accurate) then the capillary should be attached to a plastic ruler. Most plastic rulers will not melt at 100°C.

Cleaning Up

Dispose of the glass melting point capillary tubes in the container provided.

Name _____ Section _____

Lab Instructor _____ Date _____

EXPERIMENT 17 Charles' Law: Temperature/Volume Relationships for Gases

PRELABORATORY QUESTION

1. Not all melting point capillaries are the same diameter. Will a larger diameter capillary give a different value for absolute zero in this experiment? Explain.

DATA TABLE

Temperature   Length of Air Column (degree marks on thermometer or mm on ruler)

1. _____   _____

2. _____   _____

3. _____   _____

4. _____   _____

5. _____   _____

6. _____   _____

Graph of temperature versus gas volume: On a piece of graph paper make a rough plot of your six data points with the length of the air column on the *y*-axis (vertical) and the temperature, in degrees Celsius, on the *x*-axis (horizontal). Extrapolate the line to the point where the volume of the gas as measured by the length of the column is zero. The temperature at that point will be absolute zero. Once you see how the points are spaced, replot your data carefully on good graph paper so that the two scales, volume of air (column length) and temperature, fill the graph. Title the graph, label the axes, including units, and mark the six points with small circles. Draw the "best" straight line through the six points and mark the *x*-intercept.

Value of *x*-intercept when the volume of gas is zero _____

Actual value of absolute zero (±1°C) _____

Experiment 17   Report Sheet

CONCLUSION

1. What are the probable reasons for the difference between the reported value of absolute zero and your experimentally determined value?

2. Explain why, in terms of the kinetic molecular theory, the volume of air trapped in the capillary decreases as the gas cools.

3. What assumption(s) of the kinetic molecular theory explain why all gases behave alike in experiments such as this?

EXPERIMENT 18

Determination of the Molar Volume of a Gas and the Universal Gas Constant

Introduction

In this experiment you will determine the volume a mole of gas occupies. The gas is generated by allowing a known mass (and thus a known number of moles) of magnesium to react with hydrochloric acid according to this equation:

$$Mg(s) + 2\ HCl(aq) \rightarrow MgCl_2(aq) + H_2(g)$$

By measuring the volume of gas generated, the molar volume of a gas can be calculated. The kinetic molecular theory and the combined gas law suggest that we should then be able to determine a value for the universal gas constant, R, as used in the ideal gas equation of state:

$$PV = nRT$$

The hydrogen generated will be collected by the downward displacement of water in a 10-mL graduated cylinder. Because of the small amounts of material used, care is essential to justify three significant figure precision. To generate 10-mL of hydrogen requires about 9 mg (0.009 g) of magnesium. Clearly, unless you have access to an extraordinarily sensitive (and expensive) balance you will not be able to weigh the magnesium to three significant figures. This problem is solved by using magnesium ribbon, which is of uniform width and thickness.

If 100.0 cm of ribbon is weighed to three significant figures and we then use exactly 1.00 cm of the ribbon, we will know the mass of the small piece of ribbon to three significant figures.

Procedure Summary

A 1-cm piece of magnesium of known calculated weight is allowed to react with hydrochloric acid to generate hydrogen. Knowing the atomic weight of magnesium and the temperature and volume of the gas collected, the molar volume of hydrogen is calculated as well as the universal gas constant.

138   Experiment 18   Determination of the Molar Volume of a Gas and the Universal Gas Constant

Prelaboratory Assignment

Read the Introduction and Procedure sections carefully, and answer the Prelaboratory Questions on the Report Sheet.

Materials

Apparatus

Vernier caliper
10.0-mL graduated cylinder
1-hole #00 rubber stopper
22-gauge copper wire, 10 to 15 cm
Barometer
Thermometer
Wash bottle

Reagents

Magnesium ribbon, 0.9 to 1 cm lengths cleaned recently with steel wool
3 M hydrochloric acid
Sodium bicarbonate, for neutralizing acid after each trial

Safety Information
1. **Safety goggles must be worn at all times while in the laboratory.**
2. **Wear a laboratory apron.**
3. **Handle hydrochloric acid with care.** It is very corrosive. Wipe up spills, even of a single drop, immediately with a damp sponge.

Procedure

1. Fill a 400-mL beaker about three-fourths full with tap water. If possible the water should be at or near room temperature. Obtain a short (0.80- to 1.00-cm) piece of magnesium ribbon, then measure and record its length to the nearest 0.01 cm. There is a centimeter ruler in the Appendix that can be used unless you are supplied with a more precise ruler or caliper.

2. Use a 10-cm piece of copper wire to make a cage for your magnesium, by folding the magnesium over the wire, then rolling the wire around the magnesium. Dilute hydrochloric acid does not react with copper. Fit the wire cage into a 1-hole, #00 rubber stopper.

The cage should be about 2 to 3 cm from the small end of the stopper. Bend the wire over the wide end of the stopper to hold the cage in place (Figure 18.1).

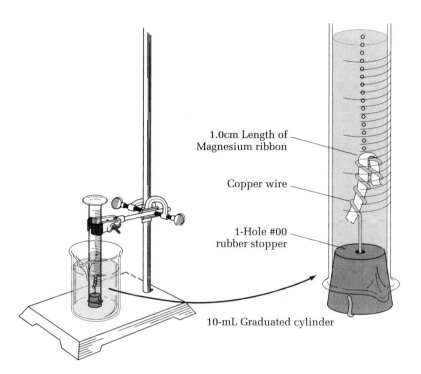

Figure 18.1 Apparatus for collecting gas by the downward displacement of water.

3. Wash a 10-mL graduated cylinder with detergent so that it drains cleanly, with no drops on the inside, then carefully pour about 3 mL of 3 M hydrochloric acid into the cylinder.

Caution! Hydrochloric acid is corrosive. Do not allow it to get on your skin or the desk! Clean up spills immediately.

4. Using a plastic wash bottle, carefully add water to the graduated cylinder until it is completely full. Try to direct the water down the side of the graduate to prevent mixing of the water and the acid, which is more dense. Insert the stopper assembly into the top of the graduated cylinder. Water should escape through the hole in the stopper; if it does not, carefully add more water, then replace the stopper. This is to ensure that no air will be trapped in the cylinder. Place your finger over the hole in the stopper and invert the graduated cylinder, lowering it into the beaker of water. Remove your finger from the stopper when it is below the level of water in the beaker.

5. Observe and record evidence of reaction. When the reaction is complete, allow the system to stand for 2 to 3 minutes, tapping the sides of the cylinder to dislodge any gas bubbles that may be clinging to the glass wall. Check also to make sure that there are no little pieces of magnesium that failed to react. Dilute hydrochloric acid does not react with copper.

6. Adjust the position of the graduated cylinder so that the water levels inside and out are nearly even to ensure that the total pressure inside the graduate is the same as the barometric pressure. Record the gas volume to the nearest 0.01 mL. Record the temperature of the water near the mouth of the cylinder. This is assumed to be the same as the temperature of the gas.
7. Repeat the experiment twice more, rinsing the cylinder, beaker, wire cage, and stopper thoroughly between trials.

Cleaning Up

After each trial neutralize the hydrochloric acid with sodium bicarbonate and flush the resulting solution down the drain. See Prelaboratory Question 2.

EXPERIMENT 18 Determination of the Molar Volume of a Gas and the Universal Gas Constant

PRELABORATORY QUESTIONS

1. In this experiment account is taken of the vapor pressure of water. What if this were not done?
   a. How would the value you calculate for the molar volume of hydrogen be affected? Explain.

   b. Would the value you calculated for R be affected? Explain.

2. In this experiment magnesium ions and chloride ions are produced. Neither is harmful to the environment. However, an excess of hydrochloric acid was used, and this cannot be poured down the drain. The hydrochloric acid can be neutralized according to the following equation

   $$HCl(aq) + NaHCO_3(s) \rightarrow H_2O + CO_2(g) + NaCl(aq)$$

   The resulting sodium chloride solution can then be flushed down the drain. Calculate the mass of sodium bicarbonate needed to neutralize 3 mL of 3 M hydrochloric acid.

Experiment 18   Report Sheet

DATA AND CALCULATIONS

Show your work on each calculation for Trial 1. Report only the answers for Trials 2 and 3.

1. Calculate the mass of the magnesium strip by using the mass of 100.0 cm of the ribbon.

2. Because the hydrogen was collected over water, the partial pressure of water vapor in the cylinder must be subtracted from the total pressure, which is equal to the barometric pressure. Measure the barometric pressure and subtract from it the vapor pressure of the water at the temperature recorded in the beaker. (See Table 6 in the Appendix.) If your barometer reads in millimeters of mercury (mm Hg), you can convert the reading to kilopascals (kPa):

   760 mm Hg = 101.325 kPa = 1 atm
   1 mm Hg = 0.1333 kPa

3. Calculate the number of moles of magnesium used in each trial (show calculations).

4. Convert water temperatures from degrees Celsius (°C) to Kelvin (K). Note that "degrees Kelvin (°K)" is incorrect nomenclature (show calculations).

CONCLUSIONS

1. Atmospheric pressure _____ mm Hg

   Convert to kPa: 1 mm Hg = 0.1333 kPa _____ kPa

Experiment 18   Determination of the Molar Volume of a Gas and the Universal Gas Constant

2. Temperature of water in beaker _____ °C

   Convert to Kelvin: 0°C = 273.12 K _____ K

3. Vapor pressure of water at measured the indicated temp. See Table 6 in the Appendix _____ kPa

4. Corrected pressure of gas in cylinder
   (atmospheric pressure minus vapor pressure of water) _____ kPa

5. Volume of gas (note units) _____ L

6. Mass of 100.0 cm of Mg ribbon _____ g

   Length of Mg ribbon reacted _____ cm

   Mass of Mg ribbon reacted _____ g

   Moles of Mg used _____ mole

   Show calculation

   Moles of H_2 produced _____ mole

For calculations 7 and 8, calculate values for each trial individually, then report an average value, based on all three trials. Include the average deviations as a part of your average values.

7. Ratio of volume of gas generated to moles of gas produced, V/n Trial 1 _____ L/mol

   Show calculation for Trial 1   Trial 2 _____ L/mol

   Trial 3 _____ L/mol

   Avg _____ L/mol

8. The universal gas constant, $R = PV/nT$ Trial 1 _____ kPaV/nT

   Show calculation for Trial 1   Trial 2 _____ kPaV/nT

   Trial 3 _____ kPaV/nT

   Avg _____ kPaV/nT

POSTLABORATORY QUESTIONS

1. Which calculation in the preceding section represents the molar volume of the gas? Explain.

2. Based on this experiment, and allowing for variations in pressure and temperature from one trial to the next, does molar volume appear to be constant?

3. Use your calculated value of R to determine the volume that 1 mole of hydrogen gas would occupy at 0°C and 101.3 kPa. This is the molar volume of hydrogen at STP.

4. The accepted value for the molar volume of hydrogen at standard temperature and pressure, (STP), is 22.43 L/mol. Calculate the percentage error for your answer to Question 3.

5. Cite possible sources of experimental error in this experiment. Remember that incorrect calculations, misuse of equipment, and poor technique are not experimental errors, hence should not be included here. Among other things consider the way in which you determined the mass of the magnesium.

6. What happened to the other product of the reaction?

EXPERIMENT 19

Oxidation-Reduction Reactions

Introduction

Many reactions involve transfer of electrons from one atom or ion to another. Such processes are known as oxidations (for electron loss), or reductions (for electron gain). Since they must always occur together, the process is often referred to collectively as "redox reactions". Many familiar types of reactions are in fact redox reactions, as you will find from the following examples.

Procedure Summary

You will observe a number of redox reactions. Based on your observations, you will be asked to draw conclusions about what is happening. Then, in Part IV, you will try to determine the relative strengths of some oxidizing and reducing agents. The list below shows six equations, called half-reactions because each shows only half of an oxidation-reduction process. Each is written first as a reduction then reversed to become an oxidation.

Reduction	Oxidation
$Zn^{2+} + 2e^- \rightarrow Zn$	$Zn \rightarrow Zn^{2+} + 2e^-$
$Cu^{2+} + 2e^- \rightarrow Cu$	$Cu \rightarrow Cu^{2+} + 2e^-$
$Mg^{2+} + 2e^- \rightarrow Mg$	$Mg \rightarrow Mg^{2+} + 2e^-$
$Fe^{3+} + 1e^- \rightarrow Fe^{2+}$	$Fe^{2+} \rightarrow Fe^{3+} + 1e^-$
$2H^+ + 2e^- \rightarrow H_2$	$H_2 \rightarrow 2H^+ + 2e^-$
$Cl_2 + 2e^- \rightarrow 2Cl^-$	$2Cl^- \rightarrow Cl_2 + 2e^-$

The experiment is in five parts, each of which illustrates a different facet of redox systems. Do the five in sequence, observing carefully all changes that occur, and record those observations in the Data Table on the Report Sheet. When numbers of drops to be used are specified, try to follow those instructions exactly; if you do not, you may miss important observation opportunities.

146   Experiment 19   Oxidation-Reduction Reactions

Prelaboratory Assignment

Read the Introduction and Procedure; answer Prelaboratory Questions 1–3 on the Report Sheet before you come to the laboratory.

Safety Information

1. **Wear safety goggles at all times in the laboratory.** A laboratory apron is strongly recommended.
2. Hydrochloric acid is corrosive to skin and clothing; handle it with care. If it is spilled on your person, wash the affected area with large amounts of water. If possible, remove clothing on which the acid has been spilled.
3. Keep $KMnO_4$ off your skin and clothes; in addition to being a strong oxidizer, it will stain anything it comes in contact with.
4. Many of the reagents used in this experiment are toxic; wash your hands thoroughly with soap and water before leaving the laboratory.

Materials

Apparatus

96-well test plate or well strips
hand lens (optional)
tweezers for handling metal pieces
safety goggles

Reagents

Reagent set, containing reagent solutions for Parts I–V, in microtip pipettes
Zinc metal pieces, about 2-mm square
Copper metal pieces, about 2-mm square
Distilled water in wash bottle
1.0 M $Cu(NO_3)_2$
1.0 M $Zn(NO_3)_2$
0.10 M $FeCl_3$
0.10 M $FeSO_4$
0.10 M $KMnO_4$
0.10 M KIO_3
0.50 M KI
3 M H_2SO_4
0.10 M NH_4VO_3
0.10 M $VOSO_4$
0.10 M $CuSO_4$

0.10 M Na_2SO_4
3% H_2O_2
$Cl_2(aq)$ - saturated
Bleach, 5.25% sodium hypochlorite, NaOCl

Procedure

Part I

1. Place 10 drops of 1.0 M HCl in a well of your test plate (Figure 19.1). Add one small piece of magnesium then observe what happens as the reaction proceeds. Record your observations in the Data Table on the Report Sheet.

Part II

2. Add 10 drops of 1.0 M copper nitrate, $Cu(NO_3)_2$, from a reagent bottle (Figure 19.2) to a clean well of your test plate. Place a small piece of zinc metal in the well with the copper nitrate. Record your observations, both of the appearance of the metal and of the solution color, in the Data Table. Changes may not happen right away.

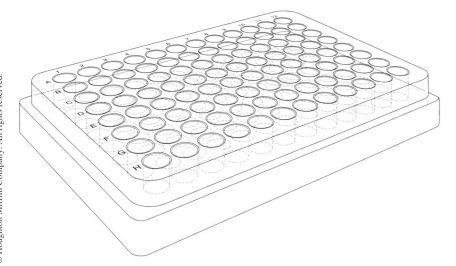

Figure 19.1 96-Well test plate.

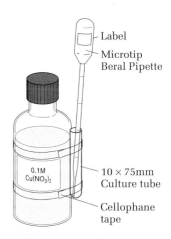

Figure 19.2 Reagent dispensing bottle.

3. Put 10 drops of 1.0 M zinc nitrate, $Zn(NO_3)_2$, in a different well of your test plate. Add a small piece of copper metal. Observe and record any reactions which take place, including solution color and changes at the surface of the metal.

Part III

4. Place 10 drops of 0.1 M iron(III) chloride (a source of Fe^{3+} ion) in one of two wells in your test plate. In an adjacent well, place 10 drops of 0.1 M iron(II) sulfate (a source of Fe^{2+} ion). Add one drop of 0.1 M potassium permanganate solution, $KMnO_4$, to each of the wells. Record your observations in the Data Table.
5. To the well in which a color change was seen, continue adding the permanganate solution, one drop at a time, until the purple color no longer changes. Record the number of drops needed to complete the reaction.

Part IV

6. Mix 2 drops of each of 0.1 M KIO_3 and 0.5 M KI together in a well. Stir with a plastic toothpick. Add 5 drops of 3 M H_2SO_4. Observe the reaction and record the results in the Data Table.

Part V

This part of the experiment requires you to make some judgments regarding the relative abilities of several ionic species to act as oxidizing agents. Your decisions will be based on a series of tests and will require a certain amount of inference on your part; you will also need to draw on some of the observations made in earlier parts of this experiment.

7. Observe and record the color of 0.1 M ammonium vanadate solution. The color is due to the presence of VO_2^+ ion, known as dioxovanadium(V), which contains vanadium(V), vanadium in the 5+ state. Likewise, observe and record the appearance of vanadyl sulfate solution. Here, the color is due to the vanadyl ion VO^{2+}, which contains vanadium(IV), vanadium in the 4+ state.

    As is the case for many transition metals, a change in vanadium's oxidation state is accompanied by a change in the color it gives to aqueous solutions. This makes it a simple matter to tell when an oxidation or reduction of the metal occurs. In your observations, be sure you associate the color you see with the name and formula of the ion which is responsible for that color.

8. Put four drops of the vanadyl (blue) solution in each of 10 wells on the top row of your clean test plate. To each of the wells 2–9, add 4 drops of one of the test solutions listed below. Add *one drop at a time*, noting any *changes that result from addition of each drop*. (The end wells are left for comparison purposes.)

Experiment 19   Oxidation-Reduction Reactions   149

Well Number	Test Solution	Species Being Tested	Well Number	Test Solution	Species Being Tested
2	$KMnO_4$	MnO_4^-	3	$FeCl_3$	Fe^{3+}
4	bleach	OCl^-	5	Na_2SO_4	SO_4^{2-}
6	$CuSO_4$	Cu^{2+}	7	3% H_2O_2	H_2O_2
8	$Cl_2(aq)$	Cl_2	9	KIO_3	IO_3^-

Observe and record any reactions which occur. Since all of the combinations involve at least one colored species, you are looking for a color change. If you don't see any change, write N.V.R. (for "no visible reaction"); if you're not sure, try putting 4 drops of the test solution in an empty well, or in a well that contains only 4 drops of deionized water.

9. Repeat the tests, using iron(II) ion, Fe^{2+}, in place of the vanadyl ion solution. Place 4 drops of 0.1 M $FeSO_4$ in each of 8 clean wells (just as you did with the vanadyl solution in Part V, Number 8), and proceed as before EXCEPT that you are to substitute ammonium vanadate (a source of VO_2^+) for the $FeCl_3$ test solution in the table.

Cleaning Up

Begin by removing to the appropriate waste container any bits of unreacted metal that may remain in your test plate. Flush remaining solutions down the drain with large amounts of water; while some of the solutes are potential hazards and others were quite acidic, the dilution provided by a running stream of water will bring the concentrations to acceptable levels.

Name _____ Section _____

Lab Instructor _____ Date _____

EXPERIMENT 19 Oxidation-Reduction Reactions

PRELABORATORY QUESTIONS

1. Define the terms *oxidizing agent* and *reducing agent*. Consult your text, if necessary. In a redox reaction, what happens to the oxidizing agent? What happens to the reducing agent?

2. What is special about the class of redox reactions known as *disproportionation reactions*?

3. Examine the lists of reagents involved in this experiment. What region of the periodic table appears to be most heavily represented?

Data Table

Part I

1.

Part II

2.
3.

Part III

4.
5. _____ drops needed

Part IV

6.

Part V

7. VO_2^+ _____ VO^{2+} _____

8. Well No./Observations   Well No./Observations

   2: _____   3: _____

   4: _____   5: _____

   6: _____   7: _____

   8: _____   9: _____

9. Well No./Observations   Well No./Observations

   2: _____   3: _____

   4: _____   5: _____

   6: _____   7: _____

   8: _____   9: _____

CONCLUSIONS

Part I

1. The only possible oxidation states for magnesium are the neutral element (zero charge) and the dipositive ion, Mg^{2+}.

   a. With which did you begin? _____

   b. What was the state of magnesium after it had reacted with HCl? _____

   c. Was magnesium oxidized or reduced? Explain.

   d. Write the half-reaction showing the change for magnesium.

Experiment 19   Report Sheet   153

2. The hydrochloric acid solution contains H⁺ ions and Cl⁻ ions. Consult the list of half-reactions given in the Introduction to decide what gas was bubbling out of the reaction. (Remember that you need one oxidation and one reduction taking place, in order to have a complete system.) Write the equation for the half-reaction resulting in the evolution of the gas you select.

3. Combine your half-reactions from questions 1 and 2, above, to write the equation for the complete reaction.

Part II

4. Consider the two wells used for zinc and copper.

   a. What evidence of reaction did you see in each case, if any?

   b. One of the following equations represents the reaction that occurred; select the correct one and justify your choice.

   $$Cu + Zn^{2+} \rightarrow Cu^{2+} + Zn \quad \text{or} \quad Cu^{2+} + Zn \rightarrow Cu + Zn^{2+}$$

   c. For the reaction equation you selected in 4b, identify: the species that is reduced; the species that is oxidized; the reducing agent; and the oxidizing agent.

Part III

5. In which well was the color of permanganate ion lost? That is, which test reagent, Fe^{2+} or Fe^{3+}, caused the purple color to fade? _____

6. The permanganate and iron solutions were all 0.1 M, yet 10 drops of iron solution required significantly less than 10 drops of permanganate. What can you conclude from these data?

7. The color change was the result of one of the iron ions being converted to the other form.

   a. Which equation best illustrates what happened in the well for which you observed the color change? Justify your choice.

   $$Fe^{2+} \rightarrow Fe^{3+} + e^- \quad \text{or} \quad Fe^{3+} + e^- \rightarrow Fe^{2+}$$

   b. Is iron being oxidized or reduced? Explain.

   c. Is permanganate ion being oxidized or reduced? Explain.

   d. Identify the oxidizing agent and the reducing agent in the reaction between permanganate and iron ions.

Part IV

8. This was a type of a disproportion reaction; the product of each half-reaction is molecular iodine, I_2. Write the two half-reactions. Decide how many electrons are gained and lost in the two half-reactions and include them in your equations. The reaction takes place in the presence of acid, so you can add water and hydrogen ions (protons) as needed to balance the numbers of atoms. Which is the oxidation and which is the reduction?

Part V

9. Which of the test ions were able to oxidize VO^{2+} to VO_2^+ (i.e., which ion(s) from the list can change vanadium from the 4+ state to the 5+)?

10. a. Which of the test ions were able to change iron(II) to iron(III)?

    b. Why was VO_2^+ substituted for Fe^{3+} as a test ion in Step 9?

    c. Of Fe^{2+} and VO_2^+, which is more easily oxidized? Defend your choice.

11. Write balanced equations for the reactions which you saw occur in Part V. Remember the following rules as guides:
    i. You must have one oxidation and one reduction in each reaction system.
    ii. Any of the reduction half-reactions which appear in the listing below may be reversed to give an oxidation.
    iii. The number of electrons lost during oxidation must equal the number of electrons gained during the reduction. To accomplish this, you will have to multiply certain of the half reactions by some factor to balance the numbers of electrons gained and lost.

    List of Reduction Half-reactions for Question 11:

    $MnO_4^- + 8H^+ + 5e^- \rightarrow Mn^{2+} + 4H_2O$

    $Fe^{3+} + 1e^- \rightarrow Fe^{2+}$

    $H_2O_2 + 2H^+ + 2e^- \rightarrow 2H_2O$

    $ClO_3^{2-} + 6H^+ + 6e^- \rightarrow Cl^- + 3H_2O$

    $HClO + H^+ + 2e^- \rightarrow Cl^- + H_2O$

    $SO_4^{2-} + 2H^+ + 2e^- \rightarrow SO_3^{2-} + H_2O$

    $Cu^{2+} + 2e^- \rightarrow Cu$

    $Cl_2 + 2e^- \rightarrow 2Cl^-$

    $VO_2^+ + 2H^+ + e^- \rightarrow VO^{2+} + H_2O$

EXPERIMENT 20

Analysis By Oxidation–Reduction Titration

Introduction

Probably the most common traditional method used for quantitative analysis of a sample is titration in aqueous solution. Titration involves measured addition of a solution of a substance whose concentration is precisely known (the *titrant*) to a sample of the material to be analyzed (the *analyte*). Traditionally, the titrant is added to the analyte by means of a buret; addition is stopped as soon as the *equivalence point* or *end point* is detected. In the familiar acid–base titration, this occurs when the concentrations of hydrogen ions (protons) and hydroxide ions are equal, and is usually detected by using an indicator—a compound that is one color in an acid and a different color in a base.

Oxidation–reduction titrations (hereafter referred to simply as *redox titrations*) are not quite so straightforward; unlike acid–base systems, for which the same type of net process is always involved, each redox system involves a unique net ionic reaction and, hence, a unique stoichiometry. Naturally, in all redox reactions the number of electrons lost in oxidation must equal the number of electrons gained in the reduction, but the number of electrons gained or lost not only varies from one species to another, it can even be different when the same substances react under different conditions! Consider the dichromate ion, $Cr_2O_7^{2-}$, with chromium in the 6+ oxidation state. Reduction can convert the chromium to any of three possible lower oxidation states: Cr^{3+}, as in $CrCl_3$; Cr^{2+}, as in $CrSO_4$; or Cr^0, chromium metal. Conversely, metallic chromium may be oxidized by varying the oxidizing agent and/or the reaction medium to a 2+, 3+, or 6+ state. In similar fashion, copper metal reacts with nitric acid to produce copper(II) nitrate. The oxidizing agent is the nitrogen in the nitrate ion of the nitric acid. Depending on the concentration of the acid, the nitrogen may be reduced to either the 2+ state (NO) or the 4+ state (NO_2).

Permanganate ion, MnO_4^-, reduces to Mn^{2+} in acid, but becomes MnO_2 when reduced in an alkaline medium. A primary difficulty encountered when designing a reaction scheme is to decide just what reagents and conditions are necessary to obtain the desired products.

On the positive side, redox titrations often do not require a separate indicator, since the color of one or more participants may change as the oxidation state of one of its elements is changed. Some examples: chromium varies from violet to red, orange, and yellow, depending on its oxidation state and the pH of the system; I^- is colorless in water and insoluble in the organic solvent, *tert*-butyl methyl ether (TBME), while I_2 is yellow in water (due to presence of the triiodide ion, I_3^-) and violet in TBME. Manganese(VII) is an inky-dark purple color as the permanganate ion, MnO_4^-, while the manganese(II) ion is such a faint pink color that it appears colorless. It is the manganese change that will serve as the end point for titrations in this experiment.

In any titration, the first task is to determine the strength of the titrant as precisely as possible, since it is the titrant concentration that is the basis for all calculations. This process is called *standardization*. Careful weighing and proper use of volumetric glassware are not always enough, since the titrant may not be obtainable in high purity, or may react with dissolved gases or other materials in the water. To contend with these problems, we use a *primary standard* to standardize the titrant before it is used. The primary standard, a compound of high purity, is generally a solid that can be weighed exactly and used without concern for contamination or side reactions. The usual primary standard for use with potassium permanganate solutions is sodium oxalate, $Na_2C_2O_4$. You will be provided with a solution of sodium oxalate whose concentration is accurately known and a potassium permanganate solution whose molarity is only approximately known.

Procedure Summary

In Part 1 of the experiment you will use the primary standard oxalate solution to standardize the permanganate; because the reaction between oxalate and permanganate ions is slow at room temperature, the solutions will be heated before the titration is begun. Then, in Part 2, the just-standardized permanganate will be used to analyze an iron compound. Both parts of the experiment must be done within a reasonably short period of time, since aqueous permanganate decomposes on standing, and this decomposition is hastened by exposure to light.

Prelaboratory Assignment

Read the Introduction and Procedure sections carefully, and answer the Prelaboratory Questions on the Report Sheet.

Materials

Apparatus

Milligram balance
Microtip beral pipettes
Labels
10-mL Erlenmeyer flasks (three or more if possible)
Sand bath, hot plate, or steam bath

Reagents

3 M H_2SO_4
$KMnO_4$
Standard sodium oxalate solution
Phosphoric acid
Iron(II) salt, or iron (II) unknown solution

Experiment 20 Analysis by Oxidation–Reduction Titration

> **Safety Information**
> 1. **Safety goggles must be worn at all times in the laboratory.**
> 2. **Handle the reagents with care.** Wash your hands before leaving the laboratory.

Procedure

The following abbreviations are used in this experiment:

PERM: the potassium permanganate solution
OX: the sodium oxalate solution; your primary standard
FE: the iron(II) unknown solution
ACID: 3 M sulfuric acid solution

Obtain about 10 mL each of the PERM, OX, and ACID solutions in separate clean, labeled containers. These quantities should be enough to complete the experiment, although you can replace solutions that are consumed, as needed.

Part 1: Standardization of Potassium Permanganate by Sodium Oxalate

1. Prepare and label three microtip pipettes as: **ACID**, for 3 M H_2SO_4; **PERM** for the $KMnO_4$ that you will standardize; and **OX**, for sodium oxalate standard. Fill each from the appropriate stock solution dispenser. Record the concentration given on the dispenser for the oxalate solution. Notice that the concentration is given in units of mole $Na_2C_2O_4$/gram of OX (sodium oxalate solution).

2. Determine the individual masses of the filled **PERM** and **OX** pipettes to the nearest 0.001 g; the mass of the sulfuric acid pipette is not needed. Place between 1.0 and 1.5 g of **OX** (about one-third of the capacity of the pipette) in a clean, dry 10-mL flask, and add an equal volume (approximately) of 3 M sulfuric acid from your **ACID** pipette. Warm the flask briefly on a hot sand bath or a steam bath. The solution is hot enough to titrate when fog begins to form on the sides of the vessel.

3. Remove the heated flask from the sand bath and begin to add the potassium permanganate solution from your **PERM** pipette, agitating the titration vessel constantly, until you have a pink color that persists for at least 30 seconds. If the titration is conducted fairly rapidly, reheating may not be necessary, but it would be a good idea to rewarm the flask briefly every 30 to 45 seconds, to maintain temperature. It is important that you keep the contents of the flask well mixed throughout the titration. If drops of permanganate should collect on the sides of the tube, use a drop or two of H_2SO_4 from your **ACID** pipette to rinse them down. Record the final masses of the **PERM** and **OX** pipettes.

4. Repeat the titration twice more, for a total of three trials. You should use a fresh flask for each trial, but you will want to rinse each vessel thoroughly immediately after use to prevent formation of permanent stains on the glass. Before proceeding, determine the ratio of mass of permanganate solution used to mass of oxalate solution for each trial. These

ratios should agree to within ±1% of the average of the three. If necessary, run additional titrations until you have three that agree within acceptable limits. Your data for all trials should be included with your report.

Cleaning Up

The contents of the titration vessels may be safely poured down the drain with large amounts of water. After you finish Part 2, reduce leftover permanganate to manganese(II) before it is flushed down the drain. The reduction is accomplished by combining all remaining **PERM**, **OX**, **ACID**, and **FE** solutions; the final mixture should be colorless. Residual brown stains in the flasks can be removed with 3% hydrogen peroxide (household antiseptic).

Part 2: Determination of the Purity of a Sample of an Iron(II) Salt

You are to determine the iron content of a sample of an iron(II) salt. Your instructor may have a solution of the salt already prepared [if so skip to 3 below]. If you are to make your own solution, you will need to prepare 25.00 mL of a solution containing a mass of about 0.6 to 0.7 g of the solid unknown. A balance with milligram sensitivity should be used, since the precision with which you prepare the sample solution is critical to your success in this part of the procedure. You will need to know the density of your iron solution, so it is important that you weigh the volumetric flask when it is clean and dry and again after the solution has been prepared.

1. To begin, weigh the sample of iron salt precisely, then transfer it quantitatively to a weighed, clean 25-mL volumetric flask. Fill the flask about halfway with distilled or deionized water, swirl to dissolve the sample, then add about 3 drops of concentrated phosphoric acid, followed by sufficient 3 M H_2SO_4 to give a total volume of 25.0 mL. Cap the flask, then invert and mix the solution thoroughly.

2. Weigh flask and contents, being careful to ensure that no residual moisture clings to the outside. (Why? Give two reasons.) Determine the total mass of the iron(II) solution you have prepared. [If the solution was prepared for you, the instructor will give you the mass of iron(II) salt used and the mass of the entire solution.]

3. Carry out the titration three separate times, following the procedure given below. As with the standardization in Part 1, do additional analyses if necessary to get three that agree within ±1% of the average value, found in the same way as before. Note that heating the flask is not necessary in this part, nor is addition of an equal volume of 3 M sulfuric acid.

    Use the solution you have just prepared to fill a clean Beral pipette labeled **FE**. Weigh the **FE** pipette and contents, then place about 1.5 g of the iron solution in a clean flask. Refill your **PERM** pipette with the permanganate solution whose concentration you determined in Part 1, weigh the **PERM** pipette and contents, then titrate the iron sample as you did the sodium oxalate, but without heating. As before, you will carry out three trials, weighing both pipettes after each trial, to determine the mass of each solution used in that trial. Clean your flasks after each trial.

Complete your final cleanup, combining the **FE** with the other solutions as described earlier in the Part 1 Cleaning Up section.

Name _____   Section _____

Lab Instructor _____   Date _____

EXPERIMENT 20   Analysis by Oxidation–Reduction Titration

PRELABORATORY QUESTIONS

1. The unbalanced equation for the reaction is

    $$Fe^{2+}(aq) + MnO_4^-(aq) \rightarrow Fe^{3+}(aq) + Mn^{2+}(aq)$$

    Balance the equation (ion-electron method).

2. (a) Balance the equation for the reaction between oxalate and permanganate ions in acidic solution.

   (b) Identify the oxidizing agent and the reducing agent in the reaction.

   (c) What is the mol ratio of permanganate to oxalate in the reaction?

Experiment 20   Report Sheet

PART 1: DATA TABLE AND CALCULATION OF PERMANGANATE CONCENTRATION

	Trial 1	Trial 2	Trial 3
Initial mass of PERM pipette	_____	_____	_____
Final mass of PERM pipette	_____	_____	_____
Mass of PERM	_____	_____	_____
Initial mass of OX pipette	_____	_____	_____
Final mass of OX pipette	_____	_____	_____
Mass of OX	_____	_____	_____

Concentration: _____ mole $Na_2C_2O_4$/mL of oxalate solution

	Trial 1	Trial 2	Trial 3
Moles of oxalate delivered	_____	_____	_____
Moles of MnO_4^- present	_____	_____	_____

Concentration of permanganate solution in mol MnO_4^-/g solution: _____

ANALYSIS AND CONCLUSIONS

Calculation of Iron Content

Using the equations you balanced in the Prelaboratory Questions, calculate the mass of iron actually present in each sample and in the entire solution. Calculate the mass of iron in the original unknown solid, from which you (or your instructor) prepared the iron (II) solution. Report the results of your calculations for each individual trial as well as an average value.

Part II

Calculation of iron Content

	Trial 1	Trial 2	Trial 3
Mass of $KMnO_4$ solution used	_____	_____	_____
Moles of $KMnO_4$ used	_____	_____	_____
Moles of iron present	_____	_____	_____
Mass of iron solution used	_____	_____	_____

Experiment 20   Report Sheet   163

Concentration of iron in mol Fe²⁺/g solution   _____   _____   _____

Moles of iron in 25.00 mL solution   _____   _____   _____

Mass of iron in 25.00 mL solution   _____   _____   _____

Mass of solid used to make solution   _____   _____   _____

% Iron in unknown solid   _____   _____   _____

POSTLABORATORY QUESTIONS

1. Suppose your solid contained some iron(III) as a result of air oxidation of iron(II). How would this affect your results? Would your values for the percentage of iron come out too high or too low? Be as specific as you can.

2. In the absence of excess hydrogen ion, reduction of permanganate solution yields the brown, insoluble compound, MnO_2. If permanganate solution is allowed to cling to the walls of the test tube, brown stains develop as a result of MnO_2 formation. How is the mole ratio of MnO_4^- to Fe^{2+} affected by this change? What effect would this have on your determinations? Explain.

3. Two other methods for determining the amount of titrant could be used. One is to put the solution in a burette, say of 10-mL capacity. Another method would be to count the drops of titrant. Why are not these methods employed in this experiment?

4. The unknown solid was either iron(II) sulfate (ferrous sulfate) or ammonium iron(II) sulfate (ferrous ammonium sulfate). Both are usually sold in hydrated form: $FeSO_4 \cdot 7H_2O$ and $Fe(NH_4)_2(SO_4)_2 \cdot 12H_2O$.

    a. For each salt, what would be mass percent of iron, assuming that you prepare 25.00 mL of solution from 0.600 grams of hydrated solid?

b. Iron(II) sulfate actually forms mono-, tetra-, penta-, and hepta-hydrates. At least two (penta and hepta) occur naturally. What are the respective mass percentages of iron for $FeSO_4 \cdot 5H_2O$ and $FeSO_4 \cdot 7H_2O$? Could you distinguish between pure samples of these two hydrates by the procedure followed in this experiment? Discuss.

c. One way to avoid difficulties due to hydration would be to heat the solid in a drying oven at 105°-110°C for several hours then cool it in a desiccator before use. Suggest one reason why that might not be suitable for ferrous ammonium sulfate.

EXPERIMENT 21

Concentration of Hydrogen Peroxide

Introduction

The label on a bottle of commercial antiseptic reads "3% hydrogen peroxide." Is this accurate? Does this mean by weight or volume? The object of this experiment is to check this value. The percentage of hydrogen peroxide, H_2O_2, in the antiseptic can be determined by reacting the peroxide solution with potassium permanganate solution. In the reaction, some of the oxygen in H_2O_2 is oxidized to O_2, while the hydrogen and the rest of the oxygen become water. The manganese from the permanganate ions is converted to Mn^{2+}, manganese(II) ions; the other products are shown in the following equations. The molecular equation for the reaction is

$$2\ KMnO_4 + 5\ H_2O_2 + 3\ H_2SO_4 \rightarrow 2\ MnSO_4 + 5\ O_2 + K_2SO_4 + 8\ H_2O$$

and the net ionic equation is

$$2\ MnO_4^-(aq) + 6\ H^+(aq) + 5\ H_2O_2(aq) \rightarrow 5\ O_2(g) + 2\ Mn^{2+}(aq) + 8\ H_2O$$

To carry out this analysis we must add exactly 2 moles of permanganate for each 5 moles of hydrogen peroxide. Potassium permanganate is a deep purple, almost black, solid that gives an intensely purple colored solution of permanganate ion. The manganese(II) ion is almost colorless, as are, of course, oxygen gas and water, the other two products of the reaction. So in order to know that exactly 2 moles of permanganate, and no more, has reacted with the peroxide a solution of potassium permanganate is added dropwise until the reaction mixture has an extremely faint pink color, indicating a very slight excess of permanganate.

On a larger scale this experiment would be run as follows: A known volume of "3%" hydrogen peroxide would be pipetted into a flask and then dark purple permanganate solution would be added to the solution from a burette, a device that allows precise known volumes of solutions to be added dropwise. When the peroxide solution turns light pink the *end point* of the *titration* has been reached, the volume of permanganate added is read from the burette, and the concentration of peroxide calculated.

On a microscale it is difficult to measure volumes from a burette to a thousandth of a milliliter, but it is easy to weigh a solution to a thousandth of a gram, so in this experiment both the peroxide solution and the permanganate solution will be weighed.

To know the amount of permanganate ion added, a solution of known concentration is prepared. The dark purple solution will contain about 0.7 g of potassium permanganate dissolved in 100 mL of water. If both the permanganate and the solution are weighed carefully, we can calculate the number of moles of permanganate in each gram of solution. This can be converted to *molarity*, *M*, by assuming the solution has a density equal to that of water.

From the mass of permanganate solution used, the number of moles of permanganate ions that have reacted can be calculated and from the stoichiometry of the reaction (permanganate and peroxide react in a 2:5 mole ratio), the number of moles of peroxide that reacted can be calculated, hence the percentage of H_2O_2 in a bottle of commercial antiseptic. The effect of temperature and/or aging on the potency of the peroxide solution can also be calculated in an additional experiment.

Procedure Summary

To determine the exact peroxide concentration in commercial "3%" hydrogen peroxide we will determine the mass of permanganate needed to react with a known mass of peroxide solution. From the amount of permanganate that reacted, we can calculate the amount of peroxide in the solution and then calculate its concentration.

You will use three solutions, each in its own pipette: potassium permanganate, $KMnO_4$; hydrogen peroxide, H_2O_2; and 6 M sulfuric acid, H_2SO_4. You will need to record the masses of the permanganate and peroxide solutions, but not of the acid.

Prelaboratory Assignment

Read the Introduction and Procedure sections carefully and answer the Prelaboratory Questions on the Report Sheet.

Materials

Apparatus

Beral pipettes, thin stem
10-mL Erlenmeyer flasks or 10 × 75-mm culture tubes or 10 × 100-mm reaction tubes
pH test paper

Reagents

Commercial hydrogen peroxide
6 M Sulfuric acid in thin-stem micro Beral pipette
Potassium permanganate solution (Note concentration on stock container)
6 M sulfuric acid

Safety Information	1.	**Safety goggles must be worn at all times in the laboratory. Aprons and/or gloves are optional.**
	2.	**Potassium permanganate, KMnO₄, is a strong oxidizing agent and will react quickly with skin and clothing. Sulfuric acid also reacts quickly with skin and clothing, causing severe burns.** Wash off a spill of either solution with large amounts of water. Notify your instuctor immediately.

Procedure

You will use three solutions, each in its own pipette: potassium permanganate, $KMnO_4$; hydrogen peroxide, H_2O_2; and 6 M sulfuric acid, H_2SO_4. You will need to keep track of the masses of the permanganate and peroxide solutions, but not of the acid.

1. Begin by filling a Beral pipette with commercial hydrogen peroxide solution and label the pipette accordingly. Fill a second pipette with $KMnO_4$ solution (dark purple), taking care to keep the solution off your skin. Record the concentration of permanganate given on the stock bottle; you will need this for your calculations. Label the pipette containing the permanganate solution. Weigh both filled pipettes and record their masses to the nearest 0.001 g.

2. Add 15 drops of the hydrogen peroxide solution to a flask, followed by 3 drops of 6 M sulfuric acid. Save the pipette with the peroxide solution to weigh later. Do not allow any peroxide solution to leak from the pipette.

3. Add potassium permanganate solution dropwise to the flask, a few drops at a time, and agitate the tube to mix the solution. If a drop of the permanganate clings to the side of the test tube, use a drop of the acid solution to rinse it down into the rest of the liquid. An excess of acid solution does no harm to this reaction. Continue adding the permanganate and mixing the solutions until a faint pink color persists.

4. Determine the masses of the permanganate and peroxide pipettes at the end of the reaction, and record the values. As soon as this trial is completed, clean the flask as directed in the Cleaning Up section.

5. Repeat the experiment twice more, for a total of three trials. Enter your data in the appropriate blanks on the Report Sheet, carry out the calculations for each trial, and report an average value for the percentage of hydrogen peroxide in the antiseptic solution. Use the same pipettes, refilling them as needed; start with a clean flask each time.

Caution!	The solution in the test tube at the end of each titration is highly acidic, and must be neutralized before you dispose of it.

Cleaning Up

Pour the contents of the tube into a beaker (250 mL or larger) and add a small amount of saturated sodium carbonate solution, a mild base. The solution will foam as the basic carbonate solution reacts with the acid. The gas escaping is carbon dioxide, CO_2. Gentle swirling of the beaker speeds the neutralization. Continue adding small amounts of base until no further foaming occurs. Leave the beaker at this point until the final trial has been completed.

Once the acid from all titrations has been neutralized, test the pH of the mixture by placing a drop on a piece of pH test paper. Compare the color of the drop on the paper with the color chart that is supplied with the paper. If it shows that the pH of the mixture is between 5 and nine, pour the liquid into the container designated as manganese(II) waste. If the pH is still below five, continue the neutralization with sodium carbonate solution until a pH nearer 7 is reached. Because Mn^{2+} is somewhat toxic, the solution will be treated further before it is disposed of.

Place your permanganate and acid pipettes, with any remaining solution, in the place designated. Do not empty them into the sink or return the unused solutions to the stock bottles.

The contents of the peroxide pipette may be rinsed down the drain with water.

Further Experiments

If you are interested in further work, your instructor may suggest one or more of the following variations:

1. Find a bottle of hydrogen peroxide whose expiration date has passed. Determine the percent hydrogen peroxide present. Has the concentration decreased with time?
2. Exposure to sunlight and heat both tend to hasten the decomposition of hydrogen peroxide. Devise and carry out an experiment to measure the effects of these variables. Obtain your instructor's approval before proceeding.
3. Certain enzymes catalyze the decomposition of peroxide. Saliva contains an enzyme called *salivary amylase*. Devise and carry out an experiment to test whether this enzyme will cause hydrogen peroxide to decompose.

Name _____   Section _____

Lab Instructor _____   Date _____

EXPERIMENT 21   Concentration of Hydrogen Peroxide

PRELABORATORY QUESTIONS

1. Study the equation given in the Introduction for the reaction of peroxide with permanganate. What would be the effect on the determination of the percent concentration of the peroxide if too much permanganate solution were added during the titration?

2. Hydrogen peroxide antiseptic carries an expiration date on the bottle. What would you expect to find if you carried out this experiment using antiseptic that was six months past its expiration date?

DATA TABLE

	Trial 1	Trial 2	Trial 3
Initial mass of pipette containing "3%" hydrogen peroxide	_____	_____	_____
Final mass of pipette containing "3%" hydrogen peroxide	_____	_____	_____
Mass "3%" hydrogen peroxide solution	_____	_____	_____
Initial mass of pipette containing potassium permanganate	_____	_____	_____
Final mass of pipette containing potassium permanganate	_____	_____	_____
Mass of potassium permanganate solution	_____	_____	_____
Concentration of potassium permanganate (on label) in mol/g	_____	_____	_____
Moles of permanganate consumed in reaction	_____	_____	_____
Moles of hydrogen peroxide consumed in reaction	_____	_____	_____

Grams of hydrogen peroxide in original solution   _____   _____   _____

Percent of hydrogen peroxide in sample   _____   _____   _____
    (g H_2O_2/wt of original solution × 100)

Average weight percentage of hydrogen peroxide in sample and
average deviation for trials 1, 2, and 3   _____

CONCLUSION

1. How do your results compare with the stated concentration on the label of the bottle? If the expiration date on the bottle has passed what general conclusion can you make about the stability of 3% hydrogen peroxide over time?

POSTLABORATORY QUESTIONS

1. Commercial hydrogen peroxide contains small amounts of organic compounds that are intended to slow the breakdown of the antiseptic. If these compounds were to react with the permanganate ions used in the titration, would your calculated value for the percentage of hydrogen peroxide be too high or too low? Explain.

2. The directions for cleanup include neutralizing the excess sulfuric acid with saturated sodium carbonate solution. Write and balance the molecular and net ionic equations for the reaction between sulfuric acid and sodium carbonate in dilute aqueous solution.

EXPERIMENT 22

The Oxidation States of Nitrogen

Introduction

Hydroxylamine, NH_2OH, is simply an ammonia molecule with an –OH group (a *hydroxyl* group) in place of one of the three hydrogen atoms. It can be regarded as a hybrid of ammonia and water. In an acidic solution, one with an excess of protons (H^+), the hydroxylamine molecule adds a proton, becoming the hydroxylammonium ion, NH_3OH^+.

When a solution of hydroxylamine hydrochloride is oxidized by acidified iron(III) ion, what is the oxidation state of the nitrogen in the product? The reaction is represented below, with Greek letters in place of coefficients, and with nitrogen in brackets, indicating that we do not know its form.

Lewis dot structures for

$$\alpha\, Fe^{3+} + \beta\, NH_3OH^+ \rightarrow \delta\, Fe^{2+} + \varepsilon\, [N?]$$

Ammonia

Since nitrogen is known to form compounds with oxidation states ranging from –3 to +5, there are a number of possible products. Among these are N_2, N_2O, NO, HNO_2 (or NO_2^-), NO_2, and NO_3^-.

Your goal in this experiment is to determine which of these possibilities is the correct product. The strategy works like this. You will use ferric ion, iron(III) to oxidize a known quantity of hydroxylamine (as hydroxylamine hydrochloride). You will use a quantity of the oxidizing agent, and that quantity is well in excess of the amount needed to carry out the reaction.

Hydroxylamine

In the reaction, iron(III) is converted to iron(II); that much is known. You will use a standardized potassium permanganate solution to reoxidize the iron(II) to iron(III); this will tell you how many moles of iron reacted. Because you also know how many moles of hydroxylammonium ion were consumed, you can establish a ratio of mol NH_3OH^+: mol Fe^{2+}. This is all you need to determine the extent to which the oxidation state of nitrogen changed and, hence, which of the product possibilities is correct.

Ammonium ion

The NH_3OH^+ ion is oxidized by the iron(III) ion; Fe^{3+} is reduced to Fe^{2+} in the process. To calculate the oxidation state of nitrogen in the products, remember that the total number of electrons lost by iron(III) must equal the total gained by the nitrogen in NH_3OH^+:

Moles of electrons lost by iron = Moles of electrons gained by nitrogen

Hydroxylammonium ion

Since iron gains 1 mole of electrons for every mole of iron reduced ($Fe^{3+} + e^- \rightarrow Fe^{2+}$), the ratio of moles of NH_3OH^+ to moles of iron will be the change in nitrogen's oxidation state.

Experiment 22 The Oxidation States of Nitrogen

Procedure Summary

A known quantity of hydroxylamine hydrochloride is oxidized with an excess of iron(III) chloride to an unknown nitrogen compound that can have an oxidation state ranging from −3 to +5. The iron(II) formed in the reaction is oxidized back to iron(III) by titration with a known quantity of permanganate. From the stoiciometry of the reactions the oxidation state of the unknown nitrogen compound can be determined.

Prelaboratory Assignment

Read the Introduction and Procedure sections, and answer the Prelaboratory Questions on the Report Sheet.

Materials

Apparatus

Milligram balance
Microtip Beral pipettes (3)
Hot sand bath, hot water bath or hot plate
10-mL Erlenmeyer flasks (3)

Reagents

Iron(III) chloride solution
Hydroxylamine hydrochloride
Potassium permanganate solution
~3 M sulfuric acid solution

Safety Information

1. **Safety goggles must be worn at all times in the laboratory.** An apron is strongly recommended for this experiment.
2. **The sand bath is hot, although it will not appear to be so.**
3. **Be careful around hot water baths.** If an open flame is used to heat the hot water bath, be certain there are no flammable liquids or vapors in the area.
4. **All of the solutions in this experiment are skin and eye irritants.** The potassium permanganate solution will oxidize skin and clothing. The iron(III) has been acidified with sulfuric acid and the hydroxylamine hydrochloride solution is irritating to skin. Keep them off your skin! Be sure to rinse any spills with water.

Procedure

1. Turn the sand bath on to 30% power. (Alternatively, prepare a hot water bath using a 100-mL beaker about half full of water.) Then obtain three microtip Beral pipettes (Figure 22.1). Label one Fe^{3+}, the second MnO_4^-, and the third Hydrox, for hydroxylamine hydrochloride. Fill each pipette with the proper solution [Fe NH_2 $(SO_4)_2(aq)$, $KMnO_4$, and hydroxylamine hydrochloride], and record the concentrations of the $KMnO_4$ and hydroxylamine solutions, in mole/g of solution. (These numbers will be found on the reagent bottles.) Weigh each pipette and record its mass. Fill a fourth pipette with approximately 3 M H_2SO_4; this will be used as needed, but its mass is of no consequence.

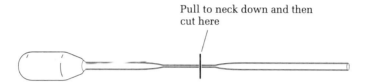

Figure 22.1 Preparation of microtip Beral pipette.

2. Place about 1 mL of the hydroxylamine hydrochloride in a 10-mL Erlenmeyer flask, then add about an equal volume (estimated) of iron(III) solution to the flask. Gently heat the mixture for about 30 seconds on the sand bath or hot plate, or in a boiling water bath for about 2 minutes. Your solution should have a slight yellow color after heating due to excess iron(III). If the solution is colorless, add a few more drops of the iron(III) solution and heat again. Cool the flask with running water for about 2 minutes, or until it has returned to room temperature.

3. Add potassium permanganate solution dropwise from the pipette to the cooled solution until a faint pink color persists. Mix the contents of the flask by swirling gently as you add the permanganate. If drops of permanganate cling to the side of your flask, try to wash them down by tilting the flask; if rinsing is needed, use 3 M sulfuric acid. Reweigh the Hydrox and MnO_4 pipettes and record the values. Clean the flask, then repeat the experiment at least two more times, or until consistent results are obtained. The Data Table in the Report Sheet contains spaces for four titrations; if additional trials are needed, continue the table on the back of the sheet. Clean your titration vessel immediately after each trial.

Cleaning Up

Dispose of all titration mixtures by flushing them down the drain; the amount of Mn^{2+} ion is too small to present a significant environmental hazard. Do not return unused quantities to the stock containers. Flasks should be washed thoroughly as soon as you finish with them, to prevent permanent stains from developing. If one or more flasks show residual stains, place a few milliliters of iron(III) solution in them and warm gently for a minute or two; this should remove the stain.

Name _____ Section _____

Lab Instructor _____ Date _____

EXPERIMENT 22   The Oxidation States of Nitrogen

PRELABORATORY QUESTIONS

1. Why is it not necessary to know the mass of the sulfuric acid pipette?

2. Assuming that hydrogen and oxygen have their usual oxidation states, what is the oxidation state of nitrogen in the hydroxylammonium ion, NH_3OH^+?

3. The experiment involves two redox reactions. Identify the oxidizing agent and the reducing agent in each reaction.

DATA TABLE

Titration 1

    Initial mass of Hydrox pipette and contents   _____

    Final mass of Hydrox pipette and contents   _____

    Initial mass of MnO_4^- pipette and contents   _____

    Final mass of MnO_4^- pipette and contents   _____

176 Experiment 22   Report Sheet

Titration 2

    Initial mass of hydrox pipette and contents _____

    Final mass of hydrox pipette and contents _____

    Initial mass of MnO_4^- pipette and contents _____

    Final mass of MnO_4^- pipette and contents _____

Titration 3

    Initial mass of hydrox pipette and contents _____

    Final mass of hydrox pipette and contents _____

    Initial mass of MnO_4^- pipette and contents _____

    Final mass of MnO_4^- pipette and contents _____

Titration 4

    Initial mass of hydrox pipette and contents _____

    Final mass of hydrox pipette and contents _____

    Initial mass of MnO_4^- pipette and contents _____

    Final mass of MnO_4^- pipette and contents _____

CALCULATIONS

Show the calculations for one of the trials in the space provided; place the results for the other trials in the blanks provided. Even though more than four trials may have been run, report only the three that show the best agreement.

1.   Calculate the number of moles of hydroxylamine hydrochloride used. [This is the same as the number of moles of hydroxylammonium ion oxidized by iron(III).]

    Trial 1: _____ mol   Trial 2: _____ mol   Trial 3: _____ mol

Experiment 22   Report Sheet   177

2. Calculate the number of moles of potassium permanganate used.

   Trial 1: _____ mol   Trial 2: _____ mol   Trial 3: _____ mol

3. Calculate the number of moles of iron(II) oxidized by permanganate.

   Trial 1: _____ mol   Trial 2: _____ mol   Trial 3: _____ mol

4. Calculate the ratio of the moles of hydroxylammonium ion to moles of iron(II) produced.

   Trial 1: _____ mol   Trial 2: _____ mol   Trial 3: _____ mol

5. Write a balanced equation for the reaction of Fe^{2+} and MnO_4^-. Remember that the reaction is taking place in an acidic medium.

6. What is the probable formula for the nitrogen-containing product? (Select the compound with nitrogen in the correct oxidation state from the list given in the Introduction.)

7. Write a balanced equation for the reaction between NH_3OH^+ and Fe^{3+} in acid solution.

Experiment 22 Report Sheet

8. Explain why it is necessary to round the ratio of hydroxylammonium ion to iron(II) ion off to the nearest whole number.

9. Suppose that as you are cooling the flask (step 2), a few drops of water enter. What effect (if any) would this have on your calculations? Explain.

10. Several possible nitrogen-containing compounds are listed in the introduction as potential products of the reaction between iron(III) ions and hydroxylammonium ions.

    a. Is hydroxylamine acting as an oxidizing agent or as a reducing agent in the reaction?

    b. Are any of the possibilities ruled out by virtue of the fact that the hydroxylamine is acting in this capacity? Explain.

    c. Refer to your answer to Prelaboratory Question 2. Is it reasonable to suggest that, under the proper reaction conditions, hydroxylamine could either be an oxidizing agent or a reducing agent? If not, why not; if so, what would have to be true of the other reagent(s) involved?

EXPERIMENT 23

Microelectrolysis of a Salt

Introduction

Electrolysis is the process of producing a chemical change by passing a direct electric current through a molten salt or a solution of a salt. The process takes place in an electrolytic cell in which positively charged ions migrate to the negative electrode and negatively charged ions migrate to the positive electrode.

Chlorine, copper, sodium hydroxide, and aluminum are all products of electrolysis on an industrial scale. When a current is passed through a concentrated aqueous solution of sodium chloride (brine) the negatively charged chloride ions (anions) will migrate to the positively charged electrode, the anode, and give up their electrons to form chlorine gas. The chloride ion has been oxidized to chlorine at the anode:

$$2\ Cl^-(aq) \rightarrow Cl_2(g) + 2\ e^-$$

The sodium ion is unaffected and does not appear in the equation.

The water at the cathode, the negatively charged electrode, will accept electrons, and become reduced to hydrogen gas and hydroxide ions:

$$2\ H_2O + 2\ e^- \rightarrow H_2(g) + OH^-(aq)$$

The overall equation is given by:

$$2\ H_2O + 2\ Cl^-(aq) \rightarrow H_2(g) + Cl_2(g) + 2\ OH^-(aq)$$

This last equation might appear to indicate that chloride ions react with water to form hydrogen gas, chlorine gas, and hydroxide ion. But this transformation will occur only by adding energy to the system in the form of an electric current. Some of this energy could be released by reacting the chlorine and hydrogen gases; they will explode.

In this experiment electrolysis will be carried out in a U-shaped tube made from glass tubing; electricity will be provided by a single 9-volt dry-cell battery.

Procedure Summary

The tube is filled with the solution to be electrolyzed, leaving about 1 cm of space near the top of each side. For Part 1, the electrolysis of potassium iodide, no electrodes are necessary; the bare ends of the wires from the battery clip may be stuck into the solution in the U-tube.

In Part 2, ordinary pencil leads will serve as electrodes. Figure 23.1 shows the U-tube with electrodes in place.

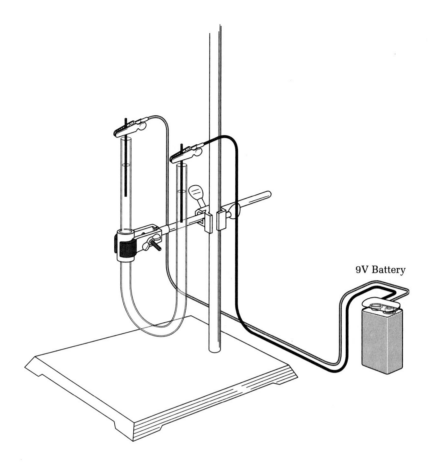

Figure 23.1
Electrolysis U-tube.

Prelaboratory Assignment

Read the Introduction and Procedure sections carefully, and answer the Prelaboratory Questions on the Report Sheet.

Materials

Apparatus

Glass U-tubes, 4–8 mm
9-V batteries
Battery clips
Alligator clips
Pencil lead electrodes

Reagents

0.5 M potassium iodide
0.5 M copper(II) chloride
0.5 M zinc bromide
Phenolphthalein indicator
Starch solution

Experiment 23   Microelectrolysis of a Salt   181

> **Safety Information**
> 1. Safety goggles must be worn at all times in the laboratory.
> 2. The reagents in this experiment are not hazardous, but use proper laboratory technique nevertheless.

Procedure

Part 1: Electrolysis of Potassium Iodide

1. A solution of 0.5 M potassium iodide is placed in the tube, leaving about half a centimeter of free space at the top. A drop of phenolphthalein solution is added to each side. A 9-V battery is connected to the battery clip and the wires from the clip are inserted directly into the solution or to a short piece of copper wire if alligator clips are at the ends of the wires. The current is allowed to flow through the solution for about 5 minutes. Note whether any gas is being evolved at either electrode. Using your hand, "waft" any gas toward your nose and carefully note any odor. Note any changes that occur during the electrolysis. Be sure to indicate whether changes are taking place at the cathode (black, negative) or anode (red, positive) side.

2. While the electrolysis is taking place add a very small crystal of iodine to a small test tube about one-quarter full of water. Note the color of the solution. Then add two drops of starch solution and note the color.

3. After you have finished observing the electrolysis, remove the wires from the solution. Then test each end of the tube with 2 drops of starch solution and note the results, if any.

Cleaning Up

Potassium iodide is not harmful to the environment so the solution can be safely poured down the drain. Rinse the electrolysis tube with water and proceed to Part 2.

Part 2: Electrolysis of Copper(II) Chloride

4. Fill the electrolysis tube with copper(II) chloride solution from the bottle. Carry out the electrolysis in the same way as described in Part 1, but eliminate the phenolphthalein indicator and use pencil leads as the electrodes. Small alligator clips soldered to the ends of the battery clip wires will help you attach the wires to the electrodes. Carefully note the polarity (positive, red; negative, black) and the appearance of the pencil lead electrodes before and after the electrolysis.

5. Note whether any gas is being evolved at either electrode. Using your hand "waft" any gas toward your nose and carefully note any odor and which electrode it comes from. Carry out this "sniff" test very warily.

Cleaning Up

Empty the copper(II) chloride solution into the container provided and rinse out the electrolysis tube with water.

Part 3: Electrolysis of Nickel Nitrate

Fill the electrolysis tube with 0.5 M nickel nitrate solution and connect the apparatus exactly as in the previous experiment, with the same pencil lead as the cathode. Carry out the electrolysis as described above. Note the appearance of the pencil lead electrodes before and after the electrolysis.

7. Note whether any gas is being evolved at either electrode. Using your hand, "waft" any gas toward your nose and carefully note any odor and which electrode it comes from. Carry out this "sniff" test very warily.

Cleaning Up

Empty the nickel nitrate solution into the container provided and rinse out the electrolysis tube with water.

Part 4: Electrolysis of Zinc Bromide

8. Carry out this part of the experiment exactly as was done for the copper(II) chloride experiment of Part 2, except fill the electrolysis tube with the zinc bromide solution from the stock bottle.
9. Note whether any gas is being evolved at either electrode. Using your hand "waft" any gas toward your nose and carefully note any odor and which electrode it comes from. Carry out this "sniff" test very warily.

Cleaning Up

Empty the zinc bromide solution into the container provided and rinse out the electrolysis tube with water.

Part 5: Electrolysis of Pure Sodium Chloride Solution and Table Salt Solution

10. Fill the tube with 0.5 M sodium chloride solution, electrolyze, and test each arm with starch solution.
11. Then fill the tube with 0.5 M table salt solution and test each arm with starch solution.

Cleaning Up

The solution can be safely poured down the drain. Rinse the electrolysis tube with water.

Name _____   Section _____

Lab Instructor _____   Date _____

EXPERIMENT 23 Microelectrolysis of a Salt

PRELABORATORY QUESTIONS

1. What two commercially valuable compounds are produced when aqueous sodium hydroxide is electrolyzed?

2. Why are graphite electrodes used in some of these experiments?

DATA AND OBSERVATIONS

Part 1

1. What happens when starch is added to a dilute solution of iodine?

2. What can you conclude when starch is added to the electrolysis tube?

3. At which electrode (positive or negative) did you get a positive test for molecular iodine? Write the equation for formation of iodine, I_2, from iodide ions, I^-. Does iodide gain or lose electrons in the process?

4. You saw bubbles of gas coming from one or both of the electrodes during the electrolysis, yet neither potassium nor iodine is a gas in its normal state.

    a. Did the gas have an odor? _____

b. What substance must have been the source of any gas that formed? (*Hint*: What was the only other substance present in the solution?)

c. i. At which electrode did you get a phenolphthalein color change? _____

ii. What ions cause phenolphthalein to change color? _____

iii. What positive ion would unite with hydroxide to form water? _____

iv. If this other ion were to gain an electron, what diatomic element would be formed? _____

v. Write the equation for formation of the diatomic element from the positive ions as the ions gain electrons. _____

d. What sort of test could you perform on the gas to help you verify its identity?

Parts 2, 3, and 4

5. Describe the evidence you found for formation of a metal from its ion in each of the electrolyses. Write equations for the gain of electrons by the metal cations to form the free metal. At which electrode did these electron-gain reactions occur in each case?

6. Chlorine gas has the odor of household bleach and bromine has a similar bad, choking odor. Describe the results of your tests for the halogens produced in each electrolysis of this part of the experiment. Write equations for the formation of the halogens from their respective halide ions. At which electrode were the halogens produced?

Experiment 23   Report Sheet   185

7. a. Add the half-reactions for formation of copper metal from copper(II) ions and for formation of chlorine gas from chloride ions. Describe in words the significance of the resulting equation.

    b. Repeat for the half-reactions for formation of zinc metal from zinc ions and formation of bromine from bromide ions. (Molecular bromine is a liquid under normal conditions.)

8. Molecular halogens are slightly soluble in water, and each lends a sort of straw yellow or orange tinge to water when in solution. Did you observe any color changes during any part of the procedure that would support the possibility that molecular halogens were formed during the electrolyses?

Part 5

9. Did you notice any difference in the electrolysis of pure sodium chloride and table salt?

10. Did testing the two electrolytes with starch show any difference? _____

11. If a difference was noted, how do you account for it? Closely examine a box of table salt for the answer.

EXPERIMENT 24

Calorimetry: Heat of Fusion of Ice, Heat of Neutralization, Specific Heat of a Metal, and Heat of Solution

Introduction

When we wish to determine the amount of heat gained or lost during a process, we use a *calorimeter* (literally, a calorie measurer) in which a thermometer measures the changes in temperature. The calorimeter is an insulated container that will allow as little heat as possible to be lost to or absorbed from the surroundings. We might also expect the container and the thermometer to absorb or release some heat energy; this amount of heat is the *heat capacity* of the calorimeter. Experimentally the heat capacity of the calorimeter is determined first, then that value is used in calculating the heat involved in all subsequent experiments done with that calorimeter. The design of the covered calorimeter is shown in Figure 24.1.

Knowledge of the magnitude of the temperature change that takes place in some process or reaction, the heat-absorbing capacity of the calorimeter, and the amount of one of the substances that is used in the process allows you to calculate the energy released or absorbed in terms of Joules (or kiloJoules) per mole of that substance. For the case where all the heat released is used to heat water, the amount of heat involved was traditionally found by the relation

$$q = m_{water} \times \Delta T \times 1 \text{ cal/g K} \tag{Eq. 24.1}$$

where the last term, 1 cal/g K, is called the *specific heat of water*; it is, in fact the definition of a calorie (the amount of heat required to heat 1 g of water 1 °C). This value for q may be changed to Joules by the conversion factor, 1 calorie = 4.184 Joules; thus, the equation which we will use is given by:

$$q = m_{water} \times \Delta T \times 4.184 \text{ J/g K} \tag{Eq. 24.2}$$

The changes in temperature can be followed in three ways, the most familiar being an ordinary mercury (or alcohol)-in-glass thermometer. A thermometer is an analog device in which the expansion of mercury corresponds to changes in temperature. Your brain converts the length of the mercury column to a digital output when you read the position of the column on the thermometer scale.

In the digital thermometer a tiny diode in the tip of the probe, powered by a battery, produces an output voltage that changes with temperature. This voltage, like the mercury in the thermometer, is an analog signal, but it can be easily converted to a digital signal by a computer chip. This signal is then read on the dial of the thermometer.

The third way in which the temperature changes in the calorimeter can be followed is through the use of a computer in which the output of a digital thermometer probe is fed into a serial port, usually the modem port or com port, of a computer. This digital signal is processed by a computer program and can be displayed on the computer screen. But the computer allows much more flexibility in the display of the data than an ordinary digital thermometer. In particular it allows the display of the temperature as a function of time. The temperature, on the

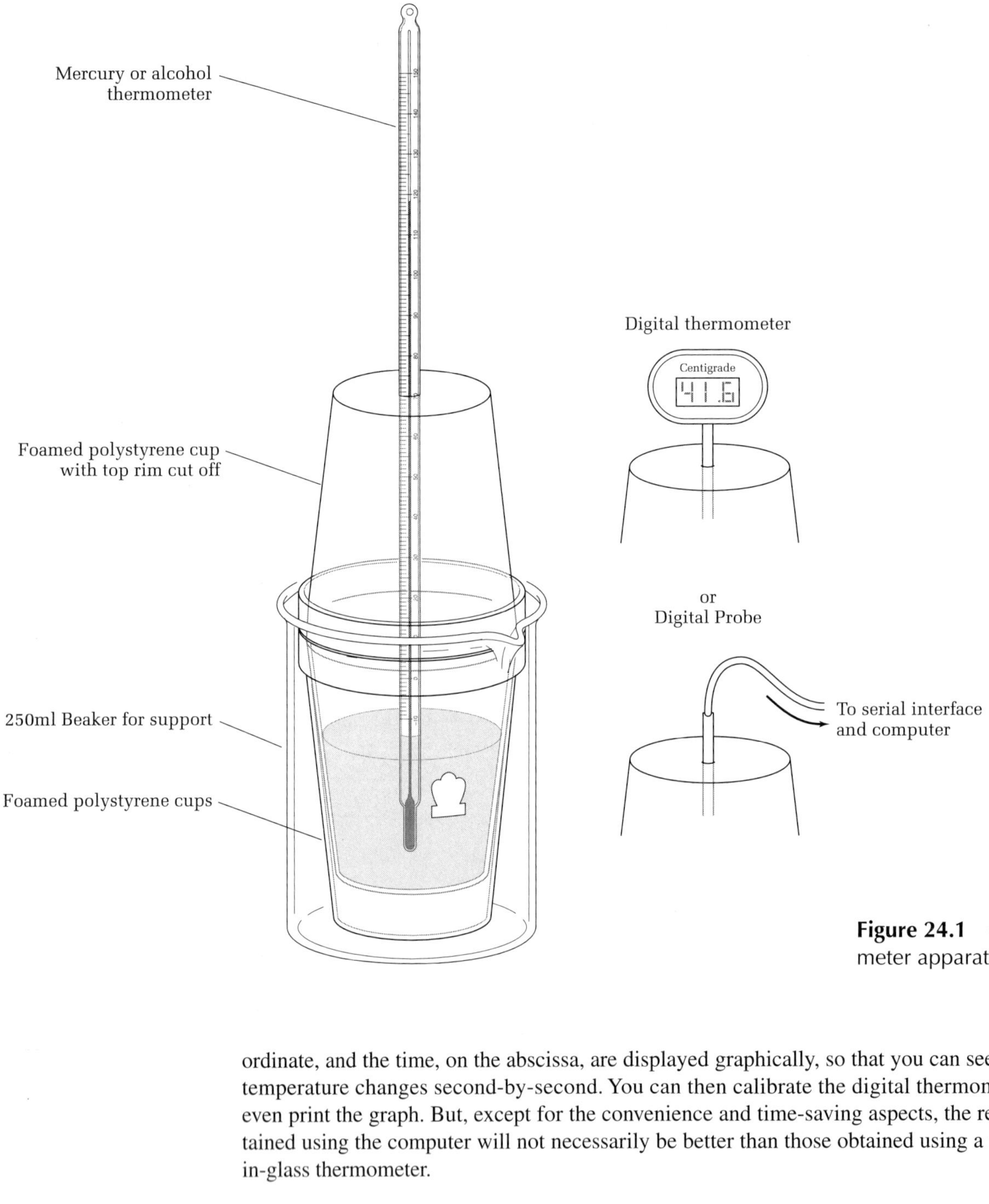

Figure 24.1 Calorimeter apparatus.

ordinate, and the time, on the abscissa, are displayed graphically, so that you can see how the temperature changes second-by-second. You can then calibrate the digital thermometer and even print the graph. But, except for the convenience and time-saving aspects, the results obtained using the computer will not necessarily be better than those obtained using a mercury-in-glass thermometer.

Procedure Summary

A calorimeter is constructed from foamed polystyrene cups. The heat capacity of the calorimeter is measured; then the calorimeter is used to determine the heat of fusion of ice, the specific heat of an unknown metal, the heat of neutralization when a strong acid reacts with a strong base, and the molar heat of solution of an unknown salt.

Prelaboratory Assignment

Read carefully the Introduction and Procedure sections and answer the Prelaboratory Questions on the Report Sheet.

Materials

Apparatus

Balance, milligram or centigram
50-mL beaker
Thermometer (mercury- or alcohol-filled, or digital, or a computer-interfaced thermometer probe)
Cap for 50-mL beaker (styrofoam)
Graduated pipette (5-mL preferred)
25 mL graduated cylinder
3 expanded polystrene cups, 6 oz.
250-mL beaker (support for calorimeter)
Hot water bath

Calorimeter Apparatus

Nest two Styrofoam cups and place a 50-mL beaker inside. Cut the top one inch from a third cup and pierce it for a thermometer or digital probe as shown in Fig. 24.1. The thermometer should touch the bottom of the beaker.

Safety Information	Safety goggles must be worn at all times in the laboratory.

Procedure

Part 1: Thermal Equilibrium and the Heat Capacity of the Calorimeter

The calorimeter used in this experiment is made of three white foamed polystyrene hot drink cups. Foamed polystyrene, as you probably already know from experience, is an excellent insulator.

1. Prepare the calorimeter by trimming the top cup just below the rim if you are using a mercury thermometer, or about 2.5 cm below the rim if you are using a shorter digital thermometer. The sharp point on the digital thermometer can be used to pierce the upper cup. Use a cork borer to make a tight-fitting hole for a mercury thermometer or a cylindrical digital probe (Fig. 24.1). If it is convenient weigh the liquids to be used in this experiment. On a top-loading balance you can easily weigh amounts to within ±0.1 g, but it is difficult using graduated cylinders to measure liquids to ±0.1 mL.

    If you are using a computer with the Vernier® or very similar Pasco® apparatus connect the digital temperature probe and the 9 V power supply into the Serial Box Interface and then connect the interface box into the modem (or printer) port on the computer. Using the Data Logger® software collect the data and process it as directed by your instructor.

2. Place exactly 25.0 mL (or 25.0 g) of *warm* water in a graduated cylinder, which you then place in a hot water bath at an accurately-known (±0.1°C) temperature between 40 and 50°C and leave for at least 5 min; this is to allow the temperature of the water in the cylinder to equal that of the bath.

3. To the dry calorimeter add 25.0 mL (or 25.0 g) of water at room temperature, determine the actual temperature (±0.1°C) of this room temperature water several times (record it) over a 4-minute period, then add all at once the 25.0 mL of warm water from the cylinder in the bath. The temperature of the bath is recorded (±0.1°C) just before the cylinder is removed and the warm water is poured into the calorimeter.

    It is important to note that the total volume of water used here will equal that used in the *Heat of Fusion* experiment that follows (Part 2). It is critical that there be no delay in transferring the hot water immediately to the room temperature water in the calorimeter; hot water will cool rapidly in air. Record the temperature (±0.1°C) every 30 seconds for about 4 or 5 minutes after mixing. Make a graph of temperature versus time (Fig. 24.2) and determine the change in temperature, ΔT, by extrapolation as seen in Fig. 24.2. Because some heat is lost through the walls of the calorimeter before the maximum temperature is reached extrapolation is necessary to determine ΔT.

Calculations

The number of Joules of heat energy gained (or lost) by water is given by the equation:

$$q = m_{water} \times \Delta T \times 4.184 \text{ J/g°C}$$

where q is the quantity of heat, m is the mass of water, and ΔT is the temperature change and 4.184 J/g°C is the heat capacity of water. In the experiment just completed, we would expect the final temperature to be exactly halfway between the initial temperatures of the hot and cold portions of water, but any absorption of heat by the calorimeter will result in a final temperature below that which is anticipated; the extent to which this happens is a measure of how much heat the calorimeter and the thermometer absorbed. It is called the heat capacity (at constant pressure) of the calorimeter, C_p, and it is calculated as follows:

C_{water} = specific heat capacity of water (4.184 J/g°C)

C_{cal} = heat capacity of the calorimeter in units of J/°C

m_{water} = mass of either the hot or room temperature water

T_H = temperature of the hot water

T_C = temperature of the cool (room temperature) water

T_F = temperature of the water at the time of mixing (obtained by extrapolation).

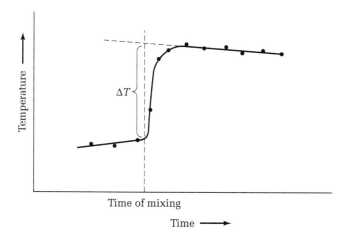

Figure 24.2 Graph of temperature versus time for calorimeter.

Then

$$C_{water}\, m_{water}\, (T_H - T_F) = C_{water}\, m_{water}\, (T_F - T_C) + C_{cal}\, (T_F - T_C)$$

The total mass of water in the calorimeter (50.0 g) is multiplied by the extent to which the final temperature fell below the predicted value, then that product is multiplied by the specific heat of water (4.18 J/g°C). This gives the quantity of heat absorbed by the calorimeter. Calculate the heat capacity of your calorimeter in J/°C by dividing the quantity of heat absorbed by the temperature change observed for the room temperature water.

Part 2: Determination of the Heat of Fusion of Ice

4. Determine the mass of the empty calorimeter assembly, add about 50 mL of water, and reweigh. Replace the thermometer and record the temperature for about 2 minutes, at which time it should become stable.

5. Add a small piece of ice (about 2 to 3 cm³) to the calorimeter. Then, quickly return the thermometer to the cup assembly and swirl gently. When the temperature has begun to rise, remove the thermometer and reweigh the calorimeter and contents to get the mass of ice that was melted. Carry out a total of three trials, drying the inside of the calorimeter with a paper towel between trials and starting with fresh water each time.

   The ice should be removed from the freezer about 15 minutes prior to running the experiment so it can "warm up" to 0°C at which point it will be wet on the outside. Wipe off the moisture from the ice before adding it to the calorimeter. Record the temperature every 30 seconds for several minutes and then plot the temperature versus time and determine ΔT as in Part 1.

Calculations

Use your data to calculate the heat of fusion of ice in J/kg. Do this with and without taking the heat capacity of the calorimeter into account. In each case, compare your experimental values with the value found using information in your textbook or a reference book; calculate your percentage error both with and without considering the heat capacity of the calorimeter system.

Part 3: Specific Heat of a Metal

Use your calorimeter to determine the specific heat of a sample of metal. Starting with 47 g of water in the calorimeter, add a sample of metal (shot) that has been in a small dry Erlenmeyer flask in a boiling water bath for at least 3 minutes. Quickly transfer the 100°C metal sample to the room temperature water; record the temperature as a function of time and determine ΔT as before. Finally determine the weight of the calorimeter plus water plus metal.

Part 4: Heat of Reaction (Heat of Neutralization)

Using 25.0 mL each of 2.00 M HCl(*aq*) and 2.00 M NaOH(*aq*), determine the heat of reaction (heat of neutralization) for the reaction

$$H^+(aq) + OH^-(aq) \rightarrow H_2O\ (l)$$

Add a drop of phenolphthalein to the acid before mixing. After mixing the solution should be pink indicating a small excess of base. If the solution is not pink quickly add NaOH solution until it is, then cap the calorimeter and record the temperature as a function of time, plot the data, and determine ΔT (Fig. 24.2).

Report your answer in kJ/mol of water; calculate your percent error.

Part 5: Molar Heat of Solution

Determine the molar heat of solution for dissolving solutes such as anhydrous calcium chloride, potassium nitrate, or ammonium chloride in water. The total volume of water in the calorimeter should be as close as possible to 50 mL in each case, to make your heat capacity for the calorimeter valid. You can assume that small amounts of solid (e.g. 0.050 mol) will make no significant difference in the volume of water, so add 0.050 mol of your solute to 50.0 mL of water.

Cleaning Up

The solutions from the calorimeter contain salts that are harmless to the environment and can be disposed of by rinsing down the drain with a large quantity of water.

Save the metal shot from Part 3 in the container provided. Do not mix one type of shot with another.

Your instructor will tell you what you should do with the polystrene cups.

Name _____   Section _____

Lab Instructor _____   Date _____

EXPERIMENT 24   Calorimetry

PRELABORATORY QUESTIONS

1. The procedure suggests using a balance to get masses of water used in the various parts of the experiment, but indicates that this is not essential. In view of the fact that the specific heat of water is given in units of joules per *gram* degree Celsius, why isn't the determination essential, rather than just helpful?

2. The British Thermal Unit, BTU, is the equivalent to the calorie in English customary units. It is defined as the quantity of heat needed to change the temperature of one pound (453.6 g) of water by one Fahrenheit degree. Calculate the Joule equivalent of one BTU.

3. A 22.50-g piece of an unknown metal is heated to 100°C then transferred quickly and without cooling into 100 mL of water at 20.0°C. The final temperature reached by the system is 26.9°C.

    a. Calculate the quantity of heat absorbed by the water.

    b. Determine the quantity of heat lost by the piece of metal.

Experiment 24 Report Sheet

c. Calculate the specific heat of the metal in J/g °C.

DATA AND OBSERVATIONS

Part 1: Heat Capacity of Calorimeter

	Trial 1	Trial 2	Trial 3
Volume or mass of room-temperature water	_____	_____	_____
Temperature of room-temperature water	_____	_____	_____
Volume or mass of hot water	_____	_____	_____
Temperature of hot water	_____	_____	_____
Average temperature of water, calculated	_____	_____	_____
Temperature of water after mixing	_____	_____	_____
Difference	_____	_____	_____

Quantity of heat absorbed by calorimeter (show calculations)

Results	_____	_____	_____

Part 2: Heat of Fusion of Ice

	Trial 1	Trial 2	Trial 3
Mass of calorimeter plus 50 mL of water	_____	_____	_____
Mass of empty calorimeter	_____	_____	_____
Mass of water	_____	_____	_____
Initial temperature of water	_____	_____	_____
Final temperature of water	_____	_____	_____
Temperature difference	_____	_____	_____

Experiment 24   Report Sheet   195

Mass of calorimeter plus water plus ice   _____   _____   _____

Mass of calorimeter plus water   _____   _____   _____

Mass of ice   _____   _____   _____

Calculation of heat of fusion of ice, taking into account the heat capacity of the calorimeter (show calculations)

Results   _____   _____   _____

Calculation of heat of fusion of ice, without taking into account the heat capacity of the calorimeter (show calculations)

Results   _____   _____   _____

Part 3: Specific Heat of an Unknown Metal

	Trial 1	Trial 2	Trial 3
Mass of calorimeter plus 47 mL of water	_____	_____	_____
Mass of empty calorimeter	_____	_____	_____
Mass of water	_____	_____	_____
Initial temperature of water	_____	_____	_____
Final temperature of water	_____	_____	_____
Temperature difference	_____	_____	_____
Mass of calorimeter plus water plus metal	_____	_____	_____
Mass of calorimeter plus water	_____	_____	_____
Mass of metal sample	_____	_____	_____

Experiment 24 Report Sheet

Calculation of specific heat of metal sample, taking into account the heat capacity of the calorimeter (show calculations)

Results _____ _____ _____

Repeat the calculation of specific heat of metal sample, but ignore the heat capacity of the calorimeter

Results _____ _____ _____

Part 4: Heat of Reaction (Heat of Neutralization)

	Trial 1	Trial 2	Trial 3
Mass of calorimeter plus 25 mL 2.00 M HCl	_____	_____	_____
Mass of empty calorimeter	_____	_____	_____
Mass of 2.00 M HCl	_____	_____	_____
Mass of beaker plus 25 mL 2.00 M NaOH	_____	_____	_____
Mass of empty beaker	_____	_____	_____
Mass of 2.00 M NaOH	_____	_____	_____
Initial temperature of acid and base (should be the same)	_____	_____	_____
Final temperature of solution after mixing	_____	_____	_____
Temperature difference	_____	_____	_____

Calculation of heat of neutralization, taking into account the heat capacity of the calorimeter (show calculations)

Results _____ _____ _____

Experiment 24   Report Sheet   197

Calculation of heat of solution of salt, taking into account the heat capacity of the calorimeter (show calculations and specify salt)

Results   _____   _____   _____

Part 5: Molar Heat of Solution

	Trial 1	Trial 2	Trial 3
Mass of calorimeter plus 50 mL of water	_____	_____	_____
Mass of empty calorimeter	_____	_____	_____
Mass of water	_____	_____	_____

Calculation of heat of neutralization, without taking into account the heat capacity of the calorimeter (show calculations)

	Trial 1	Trial 2	Trial 3
Results	_____	_____	_____
Initial temperature of water	_____	_____	_____
Final temperature of water	_____	_____	_____
Temperature difference	_____	_____	_____
Mass of calorimeter plus water plus salt	_____	_____	_____
Mass of calorimeter plus water	_____	_____	_____
Mass of salt ($CaCl_2$, KNO_3, or NH_4Cl)	_____	_____	_____

Calculation of heat of solution of salt, without taking into account the heat capacity of the calorimeter (show calculations and specify salt)

Results   _____   _____   _____

POSTLABORATORY QUESTION

1. For each of the parts of the experiment that you carried out (other than the determination of the heat capacity of the calorimeter), determine the average value of ΔH for the process and the average deviation for your three trials.

EXPERIMENT 25

Thermodynamics of Oxidation of Acetone by Hypochlorite

Introduction

In this experiment the heat of a reaction will be measured. The reaction is between sodium hypochlorite and acetone.

$$OCl^- + CH_3\overset{\overset{O}{\|}}{C}CH_3 \rightarrow CH_3\overset{\overset{O}{\|}}{C}-O^- + CHCl_3$$

This equation states that the hypochlorite ion will oxidize acetone to give the acetate ion and chloroform, trichloromethane. Hypochlorite is formed by the reaction of chlorine with hydroxide ion. A 5.25% solution is sold as household bleach and will be employed in this experiment.

The energy released during this reaction will be determined by use of a calorimeter in much the same way that a calorimeter was used in the previous experiment (Experiment 24) to measure the heat of neutralizaion of an acid with a base. See the previous experiment for a discussion of calorimetry.

Procedure Summary

A simple calorimeter is constructed and its heat capacity is measured. Then the heat of reaction of acetone with hypochlorite is determined.

Prelaboratory Assignment

Read carefully the Procedure and answer the Prelaboratory Questions on the Report Sheet.

Experiment 25   Thermodynamics of Oxidation of Acetone by Hypochlorite

Materials

Apparatus

50-mL beaker
Thermometer
Graduated pipette (5 mL preferred)
25-mL graduated cylinder
Styrofoam cups, three

Calorimeter Apparatus

Nest two Styrofoam cups and place a 50-mL beaker inside. Cut the top 1" from a third cup and pierce it for a thermometer or digital probe as shown in Fig. 25.1. The thermometer should touch the bottom of the beaker.

Reagents

Bleach
Acetone

Safety Instructions Wear your safety glasses. Bleach is irritating to the skin; wash off spilled bleach with a large quantity of water.

Procedure

Part 1: Determining the Heat Capacity of the Calorimeter

1. To a clean, dry 50-mL beaker add 12.0 mL of water at room temperature (measure it). Place the beaker and water in the calorimeter. An equal volume of water (12 mL), in a 25-mL graduated cylinder, is placed in a hot water bath at 45 to 55°C, and left for at least 5 minutes to allow the temperature of the water in the graduate to equal that of the bath.

2. A thermometer is placed directly in the graduate, and the temperature of the warm water is recorded immediately before it is poured into the beaker. It is critical that there be no delay in transferring the hot water immediately to the cool water in the calorimeter; hot water will cool rapidly in air. The thermometer in the calorimeter lid is used to record the highest temperature reached after the liquids are mixed.

Experiment 25 Thermodynamics of Oxidation of Acetone by Hypochlorite

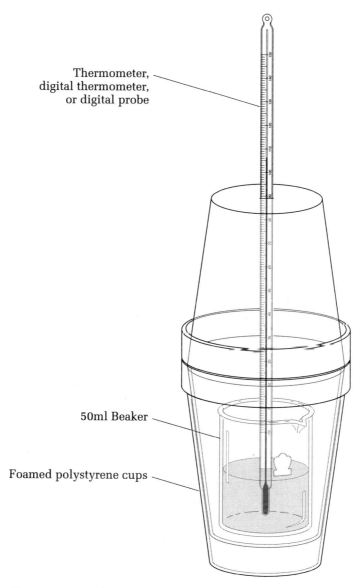

Figure 25.1 Calorimeter.

Part 2: Heat of Reaction of Acetone with Hypochlorite

3. Add 20.0 mL of household bleach [5.25% NaOCl (*aq*)] to a clean, dry 50-ml beaker in the calorimeter. Record the temperature of the bleach, which should be at room temperature. Then remove the lid and add *all at once* 4.0 mL of 5.0% aqueous acetone. Quickly replace the thermometer and lid, recording the highest temperature reached by the mixture. Swirl the beaker gently to ensure thorough mixing. The reaction is likely to take about 3 minutes to reach the maximum temperature, but the system is well enough insulated that loss of heat through the walls of the calorimeter may be neglected.

4. Rinse and dry the inside beaker and repeat the experiment at least once more; conduct additional trials as needed to obtain constant results.

Cleaning Up

Pour the contents of the beaker into the waste container provided. Each reaction produces about 0.4 g of chloroform which should not go down the drain.

Calculations

As shown in the Introduction, the number of Joules of heat energy gained (or lost) by water is given in Equation 24.2, on page 187, where q is the quantity of heat, m is mass, and ΔT is the temperature change. The data from the calibration step are used to calculate the quantity of heat absorbed by the calorimeter.

In principle, the final temperature should be exactly halfway between the initial temperatures of the hot and cold portions. However, any heat absorbed by the calorimeter will result in a final temperature below that which is anticipated; the extent to which this happens is a measure of how much heat the calorimeter absorbed, and is known as the *heat capacity of the calorimeter*, C_p. (The subscript indicates that the system is at constant pressure.)

The value of C_p is calculated as follows. The total mass of water in the calorimeter (24.0 g) is multiplied by the extent to which the final temperature fell *below* that anticipated, then by the specific heat of water (4.184 J/g K). This gives the quantity of heat absorbed by the calorimeter. To calculate the heat capacity of your calorimeter in J/K, divide the quantity of heat absorbed by the difference between the initial and final temperatures in the calorimeter from the calibration. See the previous experiment for a more extensive discussion.

The quantity of heat generated by the reaction between the bleach and acetone solutions is found in the same way, using Eq. 24.2 on page 187. Because the solutions are so dilute, they may be assumed to have the density of water as well as water's specific heat, 4.184 J/g K.

Name _____   Section _____

Lab Instructor _____   Date _____

EXPERIMENT 25 Thermodynamics of Oxidation of Acetone by Hypochlorite

PRELABORATORY QUESTIONS

1. Write the balanced equation for the reaction of chlorine with sodium hydroxide to form sodium hypochlorite.

2. Write the balanced equation for the formation of acetate ion and chloroform by oxidation of acetone with hypochlorite ion.

3. Would degrees Celsius work just as well in Eq. 24.2 on page 187? Why or why not?

DATA AND CALCULATIONS

Part 1: Heat Capacity of Calorimeter

Weight of calorimeter plus 12 mL of water _____

Weight of empty calorimeter _____

Weight of water _____

Initial temperature of water _____

Final temperature of water _____

Temperature difference _____

Weight of calorimeter plus 24 mL water   _____

Calculation of heat capacity of calorimeter

Part 2: Heat Reaction of Acetone with Hypochlorite

	Trial 1	Trial 2	Trial 3
Weight of calorimeter reactants	_____	_____	_____
Weight of empty calorimeter	_____	_____	_____
Weight of reactants	_____	_____	_____
Initial temperature of reactants	_____	_____	_____
Final temperature of reactants	_____	_____	_____
Temperature difference	_____	_____	_____
Weight of calorimeter plus weight of reactants	_____	_____	_____

Calculation of heat of reaction, taking into account the heat capacity of the calorimeter (show calculations)

CALCULATIONS

1.   a.   Calculate the *apparent* value of q for the reaction between bleach and acetone. (If you did more than one trial, calculate separate values for each one, reporting the individual values, as well as an average.)

Result   _____   _____   _____

b. For each trial done with bleach and acetone, multiply the heat capacity of your calorimeter (J/K) by the change in temperature you observed for the reaction between acetone and bleach. This gives the quantity of heat absorbed by the calorimeter for that trial.

c. To correct your apparent values of q (from part a) for the heat lost to the calorimeter, add each of your values from part b to the corresponding one from part a. This gives the actual amount of heat produced. As before, report an average value if appropriate.

2. a. Calculate the number of moles of acetone consumed in the reaction with bleach. The density of pure acetone is 0.791 g/mL, and the solution was 5.0% acetone by volume.

b. Calculate the heat of reaction for the oxidation of acetone by bleach, in kJ/mol $(CH_3)_2CO$.

3. a. Calculate the *apparent* value of q for the reaction between bleach and acetone. If you did more than one trial, calculate separate values for each one, reporting individual values, as well as an average. You need show calculations for only one trial.

b. For each trial done with bleach and acetone, multiply the heat capacity of your calorimeter (J/K) by the change in temperature you observed for the reaction between acetone and bleach. This gives the quantity of heat absorbed by the calorimeter for that trial.

c. To correct your apparent values of q (from **a**) for the heat lost to the calorimeter, add each of your values from **b** to the corresponding one from **a**. This gives the actual amount of heat produced. As before, calculate individual values for each trial, then report an average value, as well.

4. a. Calculate the number of moles of acetone consumed in the reaction with bleach. The density of pure acetone is 0.791 g/mL, and the solution was 5.0% acetone by volume.

b. Calculate the average value of the heat of reaction for the oxidation of acetone by bleach, in kJ/mol $(CH_3)_2CO$.

5. Review your calculations. How significant was the correction for the heat capacity of the calorimeter to your final result?

6. Is it possible to calculate a value for the change in internal energy, ΔE, of this system? If so, how would one go about it? If it is not possible, why not?

EXPERIMENT 26

Reaction Kinetics: The Aldol Condensation of Acetone with Benzaldehyde

Introduction

In this experiment an important organic reaction, the aldol condensation, will be carried out and the effects of temperature and concentration on the rate of the reaction will be studied.

Colorless acetone, a liquid with a boiling point of 56°C, and colorless benzaldehyde, a liquid of boiling point 180°C that smells somewhat like almonds, will react with each other in the presence of a base, sodium hydroxide, in a stepwise fashion to give a yellow crystalline product called dibenzalacetone:

$$2 \; C_6H_5CHO + CH_3COCH_3 \xrightarrow{\text{NaOH, MW 40.01}} C_6H_5CH=CHCOCH=CHC_6H_5$$

Benzaldehyde
MW 106.13, bp 178°C
den 1.04

Acetone (Propanone)
MW 58.08, bp 56°C
den 0.790

**Dibenzalacetone
(1,5-Diphenyl-1,4-pentadien-3-one)**
MW 234.30, mp 110–111°C
λ_{max} 320 nm, ϵ 34,300

When the three reactants are mixed, the solution is initially colorless and transparent. It turns faintly yellow (still transparent) and then suddenly the entire solution becomes opaque as the dibenzalacetone begins to come out of solution. The number of seconds between the addition of the sodium hydroxide and the appearance of opaqueness is what we will use to measure the rate of the reaction. Crystals of bright yellow dibenzalacetone will appear some time after the solution turns opaque.

A rough "rule of thumb" regarding organic reactions says that the rate of a reaction will approximately double or triple with each 10°C rise in temperature, and the rate of reaction will usually depend on the concentration of one or more of the reactants. In this experiment we explore the effect of these reaction conditions on the rate of reaction.

Procedure Summary

Benzaldehyde is allowed to react with acetone in the presence of a basic catalyst under a variety of conditions in order to assess the effect of temperature and concentration on the rate of chemical reaction.

Prelaboratory Assignment

Read the Introduction and Procedure sections, and answer the Prelaboratory Questions on the Report Sheet.

Materials

Apparatus

13 × 100-mm test tubes, reaction tubes, or 100-mm culture tubes
1.0-mL pipettes
Pipette pumps, rubber bulbs, or syringes to fill pipette
250- or 400-mL beakers, for water baths
Ice
Hot water or hot plates
Thermometers
Glass stirring rods
100-mL beakers, for waste
Funnels and filter paper to collect crystals
Stopwatches

Reagents

1 M acetone in ethanol
2 M benzaldehyde in ethanol
3 M sodium hydroxide
Ethanol
Ice
3 M hydrochloric acid, for use in Cleaning Up section

Safety Information

1. **Safety goggles must be worn at all times in the laboratory.**
2. **Sodium hydroxide solution is strongly caustic.** Avoid contact and wipe up spills immediately. Wash your hands thoroughly should you come in contact with sodium hydroxide. It will feel slippery on the skin.
3. **Benzaldehyde is listed as toxic irritant.** Although a dilute ethanol solution is sold in the grocery store as artificial almond flavor, do not taste it.

Procedure

Part 1: The Reaction at Room Temperature

Carry out this reaction three times at room temperature. Begin timing when the sodium hydroxide solution is added to the mixture of acetone and benzaldehyde. Both the acetone and the benzaldehyde have been dissolved in ethanol, an inert solvent in this reaction. Note the concentrations of these reactants on the bottles.

1. Into a reaction tube or 100-mm culture tube or test tube add 1.00 mL of acetone solution and 1.00 mL of benzaldehyde solution. These solutions have been prepared so that the concentration of benzaldehyde is exactly twice that of the acetone. The more accurately you measure these solutions, the more reproducible your results will be; it is best to use a 1-mL glass or plastic graduated pipette, which should be rinsed with 1 mL of ethanol between uses. Stir the solution thoroughly with a glass rod, measure the temperature of the solution, remove the thermometer, rinse it with water and dry it.

2. To start the reaction add 2 mL of the catalyst, 3 M aqueous sodium hydroxide solution. It need not be measured as accurately as the reactants. Record the time and immediately stir the reaction mixture thoroughly with a glass rod. Record the time when the solution turns opaque. Set the tube aside and repeat the reaction twice more in clean tubes.

Part 2: Effect of Temperature and Concentration on the Rate of Reaction

A. High-Temperature Reaction (35–40°C)

3. Carry out the reaction exactly as described in Part 1 but before adding the sodium hydroxide to the mixture of acetone and benzaldehyde immerse tubes containing the reagents, the catalyst, and the stirring rod in a beaker of water that is about 35 to 40°C (hot tap water or water heated on a hot plate). Three or 4 minutes should be long enough for the temperature of the solutions to come to equilibrium.

4. Note the temperature of the mixture just prior to adding the sodium hydroxide. Rinse the thermometer in the beaker of water and dry it off between experiments, which should be done a total of three times at the elevated temperature.

B. Low-Temperature Reaction (10–15°C)

5. Carry out the reaction exactly as described in Part 1 but this time put ice into the beaker to bring the temperature of the tubes containing the reagents down to between 10 and 15°C. Again carry out the reaction a total of three times.

C. Concentration of Acetone and Benzaldehyde

6. At room temperature run three reactions using just half the above quantities of acetone and benzaldehyde (0.50 mL of each). Add 1 mL of ethanol to keep the volume constant and use the same quantity of sodium hydroxide solution (2 mL). This will keep the total volume of the reaction mixture and the volume of water constant.

D. Concentration of Sodium Hydroxide

7. At room temperature run three reactions using just half the quantity of sodium hydroxide solution. Use 1.0 mL of sodium hydroxide solution, 1.00 mL of water, 1.00 mL of acetone solution, and 1.00 mL of benzaldehyde solution. This will keep the total volume of the reaction mixture and the volume of water constant. As usual stir the solution thoroughly with a glass rod, measure the temperature of the solution, remove the thermometer, rinse it with water and dry it.

Cleaning Up

Scrape and pour all of the reaction mixtures into a beaker. Collect the crystals of dibenzalacetone by filtration on a cone of filter paper, wash the crystals with a little ethanol, neutralize the filtrate with dilute hydrochloric acid, and wash the solution down the drain with lots of water. Put the dibenzalacetone crystals in the organic waste container.

Name _____ Section _____

Lab Instructor _____ Date _____

EXPERIMENT 26 Reaction Kinetics: The Aldol Condensation of Acetone with Benzaldehyde

PRELABORATORY QUESTIONS

1. From your experience outside the laboratory cite some examples of the effect of temperature on the rate of a chemical reaction.

2. Step 2 of the procedure specifies that the quantity of the catalyst, NaOH, need not be measured as accurately as those of the reactants. Why not?

CALCULATIONS

Calculate the average time for completion of the reaction at room temperature (Part 1). If two of the three runs are in significantly better agreement than the third, you may elect to average only those two, but your report should show all three, and you should explain why one of the times was discarded. The reaction rates can be expressed as the number of moles of reactant that react per some unit of time, in this case, per second.

In the same manner as for Part 1, calculate the average time to complete the overall reaction at the elevated temperature and at the reduced temperature and then the reaction rates at the two temperatures. How does temperature seem to affect the rates? How do your results compare with the "rule of thumb" that says that the rate of an organic reaction approximately doubles or triples with each 10° rise in temperature? Discuss your findings, making allowances for variations in temperature within each part.

DATA

Part 1: The Reaction at Room Temperature

Trial	Temp (°C)	Time to Opaque Appearance (seconds)
1	_____	_____
2	_____	_____
3	_____	_____

Average time for completion of reaction _____

Part 2A: High-Temperature Reaction (35–40°C)

Trial	Temp (°C)	Time to Opaque Appearance (sec)
1	_____	_____
2	_____	_____
3	_____	_____

Average time for completion of reaction _____

Part 2B: Low-Temperature Reaction (10–15°C)

Trial	Temp (°C)	Time to Opaque Appearance (sec)
1	_____	_____
2	_____	_____
3	_____	_____

Average time for completion of reaction _____

Part 2C: Concentration of Acetone and Benzaldehyde

Trial	Temp (°C)	Time to Opaque Appearance (sec)
1	_____	_____
2	_____	_____
3	_____	_____

Average time for completion of reaction _____

Part 2D: Concentration of Sodium Hydroxide

Trial	Temp (°C)	Time to Opaque Appearance (sec)
1	_____	_____
2	_____	_____
3	_____	_____

Average time for completion of reaction _____

RESULTS

1. From a knowledge of the concentrations of the acetone and benzaldehyde (on the bottles) and the total volume of the reaction mixture, calculate the initial concentrations of the two reactants.

   Initial concentration of acetone _____ mol acetone/L

   Initial concentration of benzaldehyde _____ mol benzaldehyde/L

Rates of Reaction

1. At room temperature

   _____ (mol acetone/L)/sec

   _____ (mol benzaldehyde/L)/sec

2A. At high temperature

   _____ (mol acetone/L)/sec

   _____ (mol benzaldehyde/L)/sec

2B. At low temperature

   _____ (mol acetone/L)/sec

   _____ (mol benzaldehyde/L)/sec

2C. At one-half concentration of acetone and benzaldehyde

   _____ (mol acetone/L)/sec

   _____ (mol benzaldehyde/L)/sec

2D. At one-half concentration of acetone and benzaldehyde

_____ (mol acetone/L)/sec

_____ (mol benzaldehyde/L)/sec

2E. At one-half concentration of sodium hydroxide solution

_____ (mol acetone/L)/sec

_____ (mol benzaldehyde/L)/sec

2. Are these rates a positive or a negative quantity? Explain.

3. How well do the low- and high-temperature reactions conform to the rule of thumb that says the rate of an organic reaction will double or triple with a 10°C rise in temperature?

4. Rationalize the rates of reaction that result when the concentrations of the organic reactants are halved.

5. Rationalize the rate of reaction that results when the concentration of the sodium hydroxide solution is halved.

Optional

The Arrhenius equation states in quantitative terms the relationship between the specific reaction rate constant, k, and the activation energy, E_a:

$$\ln k = \ln A + E_a/RT$$

where A is a constant, R is the gas constant, and T is the temperature in Kelvin.

1. Make a plot of time (of opaqueness) versus temperature. Describe the shape of the curve you obtain.

2. Make a second plot, showing time versus $1/T$ where T is the temperature in Kelvin. Describe the shape of the curve.

3. Plot ln time versus $1/T$. Describe the resulting curve.

4. Make plots of rate versus temperature (in K), rate versus $1/T$, and ln rate versus $1/T$. Describe the shapes of the curves.

5. What can you suggest about the order of the reaction overall?

6. The Arrhenius equation may be used to show that, for the same reaction carried out at two or more temperatures,

$$\ln (k_2/k_1) = (E_a/R)[(1/T_1) - (1/T_2)]$$

where k_1 and k_2 are the specific rate constants at the respective temperatures, T_1 and T_2; E_a is the activation energy; and R is the gas constant, 8.31×10^{-3} kJ/mol K. Show that the value of k_1/k_2 should equal the ratio of reaction times, t_2/t_1. Then calculate the value of the activation energy, E_a, using this information. Summarize and discuss your findings.

EXPERIMENT

27 Metal Reactivities

Introduction

Metals form compounds by losing electrons and becoming positive ions (*cations*). If the process is reversed, the cations regain the lost electrons and become neutral atoms once again. The *active* metals are the ones that react most readily; this means that they lose electrons easily. A metal that does not lose electrons easily is said to be *inactive*. There are degrees, of course, of activity and inactivity.

Some metals are so active that they must be kept from all contact with water or even air. This is typical of the first group (the first vertical column) on the left of the periodic table. At the other extreme, metals like gold and platinum are almost totally unreactive. All the other metallic elements fall in between these extremes, resulting in what is known as the *activity series for metals*, in which the metals are listed from most active to least active; in a typical list, number 1 is cesium, number 48 is gold.

As you can quickly deduce, not all the metals are included in any such list; in particular, radioactive elements and the very rare metals are omitted. In fact, the list is often shortened to only about 20 metals for simplicity. Some very active metals are somewhat deceiving because when exposed to air they very rapidly form a protective coat of oxide on the outside. Aluminum is the prime example.

An active metal will give up electrons to the cation form of an inactive metal. For example, zinc is active, whereas nickel is fairly inactive, so if you were to put a piece of zinc in a solution containing nickel ions, the nickel ions would take electrons from the zinc atoms and metallic nickel and a solution of zinc ions would result. In equation form:

$$Zn(s) + Ni^{2+}(aq) \rightarrow Zn^{2+}(aq) + Ni(s)$$

As you know, if the solution contains positive ions, it must also contain negative ions. Since the negative ions (*anions*) do not take part in the reaction, they are generally left out of the equation. Here, though, is the same equation shown with the anions:

$$Zn(s) + NiCl_2(aq) \rightarrow ZnCl_2(aq) + Ni(s)$$

While chloride ions (Cl$^-$) were used as the anions in this example, it is also quite common to see nitrate ions, NO_3^-, as the negative ion. These two are used more often than most others because they are the least likely to react with anything else in the solution.

To summarize, ions of less-active metals take electrons from (oxidize) neutral atoms of more-active metals. In this experiment, you will determine the relative activities of five different metals by placing samples of each metal in solutions containing the ions of the others. The metals you will test are copper, iron, magnesium, tin, and zinc. The reactivity of silver will be demonstrated during the postlab discussion.

220   Experiment 27   Metal Reactivities

Procedure Summary

In this experiment you will test the relative activities of several metals and then order the metals from most active to least active.

Prelaboratory Assignment

Read the Introduction and Procedure sections carefully, and answer the Prelaboratory Question on the Report Sheet.

Materials

Apparatus

24-well test plate
Forceps or tweezers
Wash bottle
Magnifying glass (optional)

Reagents

Microtip pipettes containing:
0.2 M $CuCl_2(aq)$
0.2 M $FeCl_3(aq)$
0.2 M $Mg(NO_3)_2(aq)$
0.2 M $SnCl_4(aq)$
0.2 M $Zn(NO_3)_2(aq)$
small pieces of Cu, Fe, Mg, Sn, Zn

Safety Information
1. **Safety goggles must be worn at all times in the laboratory.**
2. **Handle the reagents with care.** Wash your hands before leaving the laboratory.

Procedure

1. Using tweezers or forceps, place single pieces of one of the metals in four separate wells of your 24-well test plate according to the diagram shown in Figure 27.1. Cover each piece with 8 to 10 drops of a different metal solution. *Do not use the solution that contains ions of the metal being tested.* Observe carefully, because the only sign of reaction may be a darkening of the metal surface. Some reactions will not occur immediately; wait at least 8 to 10 minutes before you decide that nothing happened in order to detect slow reactions. Record evidence of reaction in the Data Table on the Report Sheet. If no reaction occurs, record "NR."

	Cu^{2+}	Fe^{2+}	Mg^{2+}	Sn^{2+}	Zn^{2+}
	Cu	Cu	Cu	Cu	Cu
	Fe	Zn	Fe	Fe	Fe
	Mg	Mg	Zn	Mg	Mg
	Sn	Sn	Zn	Zn	Sn

Figure 27.1 Placement of metals in 24-well test plate.

2. Repeat Step 1 for the four other metals. If you are uncertain about the results of any pairing, use one of the empty wells in your test plate to test the metal in question with distilled water as a control.
3. Your instructor may direct you to leave your well plate overnight in a safe place to see if any further changes occur. If so, obtain a cover for the well plate, label it, and place it in the location specified by your instructor.

Cleaning Up

Shake the contents of your test plate into the large pan lined with paper towels. Use forceps or tweezers to remove pieces of metal that remain in the wells. Then use a wash bottle to rinse solutions from the test plate onto the paper towels and shake the water from the wells. Your instructor will complete the disposal process. Before leaving the laboratory, clean up all other materials and wash your hands thoroughly.

Name _____   Section _____

Lab Instructor _____   Date _____

EXPERIMENT 27 Metal Activities

PRELABORATORY QUESTION

1. The electrical contacts of computer chips are often gold plated. Why?

DATA TABLE RESULTS AND OBSERVATIONS

Solutions of Cu^{2+}, Fe^{2+}, Mg^{2+}, Sn^{2+}, Zn^{2+} are added to pieces of metal in 20 wells. Record any changes in the appearance of the metal, or any other evidence of reaction.

	Cu^{2+}	Fe^{2+}	Mg^{2+}	Sn^{2+}	Zn^{2+}
	Zn	Cu	Cu	Cu	Cu
	Fe	Zn	Fe	Fe	Fe
	Mg	Mg	Zn	Mg	Mg
	Sn	Sn	Sn	Zn	Sn

Demonstration by instructor of reaction between copper metal and silver ions:

ANALYSIS AND CONCLUSIONS

1. Which metal reacted with the most solutions of other metal cations? (It is the most active.)

2. Of the ones you tested, which metal reacted with the fewest solutions of other metal cations? (It is the least active.)

3. Rank the five metals that you tested in order of decreasing activity (most active first).

4. Refer to your observations of the reaction between copper metal and silver ions, which was demonstrated by your instructor. Based on what you saw, assign silver its position on the activity list.

POSTLABORATORY QUESTIONS

1. Aluminum metal, Al, will react with a solution containing copper(II) ions, Cu^{2+}, but copper metal, Cu, will not react with a solution of aluminum ions, Al^{3+}. Which metal is more active, copper or aluminum? Justify your answer.

2. Iron metal, Fe, will react with a solution of barium ions, Ba^{2+}. Would you expect barium metal, Ba, to react with a solution of iron(III) ions, Fe^{3+}? Why or why not?

3. Several equations are written below. Some of them describe reactions that took place during your experiment; the rest show reactions that did not occur. Using your observations as a guide, circle the letters of those reactions that did occur, and put a line through the equations that describe reactions that did not take place. All of the equations are written correctly; all you have to do is decide which ones describe reactions that actually took place and which do not.

a. $Zn(s) + Mg(NO_3)_2(aq) \rightarrow Zn(NO_3)_2(aq) + Mg(s)$

b. $Mg(s) + CuCl_2(aq) \rightarrow MgCl_2(aq) + Cu(s)$

c. $MgCl_2(aq) + Cu(s) \rightarrow Mg(s) + CuCl_2(aq)$

d. $4\ Fe(s) + 3\ SnCl_4(aq) \rightarrow 4\ FeCl_3(aq) + 3\ Sn(s)$

e. $3\ Sn(s) + 4\ FeCl_3(aq) \rightarrow 3\ SnCl_4(aq) + 4\ Fe(s)$

f. $2\ Mg(s) + SnCl_4(aq) \rightarrow 2\ MgCl_2(aq) + Sn(s)$

g. $3\ Zn(s) + 2\ FeCl_3(aq) \rightarrow 3\ ZnCl_2(aq) + 2\ Fe(s)$

h. $Cu(s) + 2\ AgNO_3(aq) \rightarrow Cu(NO_3)_2(aq) + 2\ Ag(s)$

4. Consider the following observations:

$Ni(s) + CuCl_2(aq) \rightarrow$ reaction occurs

$Ni(s) + Mg(NO_3)_2(aq) \rightarrow$ NR

$Ni(s) + SnCl_4(aq) \rightarrow$ reaction occurs

$Ni(s) + FeCl_3(aq) \rightarrow$ NR

Use these observations to place nickel on the activities list that you wrote in the preceding Analysis and Conclusions section, Question 3.

5. Examine equations a and g in Question 3 above. In one case, zinc metal becomes zinc nitrate, $Zn(NO_3)_2(aq)$; in the other, the product is zinc chloride, $ZnCl_2(aq)$. Why does the same metal form different products in the two different reactions?

6. Complete and balance the reaction equations for the two reactions that *do* occur in Question 4.

7. Tin metal foil, such as you used in this experiment, like many metals, often has a thin coating of oil put there during the manufacturing process.
   a. Suggest a reason for this.

   b. Discuss the effects the presence of such an oil coating would have on tyhis experiment if it were not removed.

   c. If it is to be removed, would rinsing with water be adequate? Why or why not? (Hint: Refer to experiment 2, Densities of Organic Liquids or Experiment 6, Solubility and Solutions, for discussions of solvent miscibility.)

   d. Would such a coating be more necessary for elements near the start or the end of a typical activity list? Why?

EXPERIMENT 28

Nonmetal Reactivities

Introduction

Elements are described as being active if they tend to react with other elements to form compounds readily; in that sense, the word *active* really means *reactive*. In contrast to metals, which can only bond ionically, nonmetals can form compounds in either of two ways: ionic bonding or covalent bonding. In ionic bonds, nonmetals tend to gain electrons, forming anions. Although covalent bonding results from the sharing of electrons, it nonetheless involves atoms donating electrons to each other to fill available orbitals. Thus, an active nonmetal is one that has a strong tendency to acquire electrons.

In this experiment we will work with three of the four common halogens, chlorine, bromine and iodine. The halogen fluorine is so highly reactive, poisonous, and dangerous it is rarely encountered in the laboratory; hydrofluoric acid is a very hazardous substance because it gives burns that heal very slowly. In addition, the common soluble fluorides are very toxic. The fifth halogen in the periodic table, astatine, is radioactive, rare, and has never been isolated in pure form.

This experiment will provide you with a chance to compare the relative electron-attracting abilities of the common halogens by setting the molecules (Cl_2, Br_2, and I_2) in direct competition with their ions (Cl^-, Br^-, I^-). Based on tests with certain of the combinations, you should be able to infer the order of relative activities for the entire halogen family.

Procedure Summary

Solutions of the halide ions will be mixed with aqueous solutions of the gases. If a halogen is freed in the process, it will be extracted with an organic solvent and identified by color. One of the halogens, but not its ion, reacts with starch, which is a confirmatory test.

Prelaboratory Assignment

Read the Introduction and Procedure sections, and answer Prelaboratory Questions on the Report Sheet.

Materials

Apparatus

Reaction tubes or 10 × 100-mm culture tubes
Beral pipettes, graduated, 1 mL
Corks to fit tubes

Reagents

Chlorine water
Bromine water
Iodine water
tert-Butyl methyl ether, in dropping bottles
1.0 M sodium bromide
1.0 M sodium chloride
1.0 M sodium iodide
Sodium sulfite solid, 1 g

Safety Information

1. **Safety goggles must be worn at all times in the laboratory.** Contact lenses should not be worn for this or any other experiment.

2. **The halogens have strong, irritating odors, and can damage mucous tissue.** Avoid direct inhalation of these vapors. Wear gloves and dispense the solutions in the hood.

Procedure

Part 1: Preliminary Tests

1. Place 0.5 mL of TBME (*tert*-butyl methyl ether) in each of three reaction tubes or in 10 × 100-mm culture tubes. To one of the tubes add 0.5 mL of chlorine water, $Cl_2(aq)$; add 0.5 mL of iodine water, $I_2(aq)$ to the second; and place 0.5 mL of bromine water, $Br_2(aq)$, in the third. The approximate graduations on a plastic Beral pipette are good enough for these measurements.

2. Cork each of the tubes and shake briefly to mix. Note and record the appearance of the contents of each tube. Add a few drops of starch solution to each tube, shake, and note the result.

Part 2: Activity Tests

3. Into each of three separate tubes, place 0.5 mL of one of the three halide ion solutions; be sure to keep track of which halide ion is in which tube. Add 0.5 mL of TBME to each tube, followed by 0.25 mL of chlorine water. Cork the tubes and shake briefly to mix. Allow the contents to settle and record the appearance of both layers in each tube.

4. Repeat the process, this time substituting bromine water for chlorine water. Again, describe the results of the tests. If you need to reuse the same tubes, be sure to clean them as described below.

Cleaning Up

Although the halide ion solutions present no chemical hazard, both TBME and the molecular halogens must be disposed of carefully. Addition of sodium sulfite solution, followed by gentle shaking (cork) will convert the halogens to halide ions and will simultaneously move them out of the organic layer (TBME) and into the lower aqueous layer. When the layers are separated the bottom aqueous layer can go down the drain, flushed with water. The organic layer should go in the container provided.

Analysis of Results

For those combinations in which a reaction took place, write molecular and net ionic equations for the reactions that took place. You need not write equations for any cases in which there was no chemical reaction.

Examination of your data should let you place bromine, chlorine, and iodine in order of relative activity. Bear in mind that a more active molecular halogen is able to oxidize (take electrons from) less-active halide ions. Once you have established the order for these three, you should be able to place fluorine relative to the other three, based on their positions on the periodic table.

Suppose you had been provided with solutions of molecular fluorine and molecular iodine—$F_2(aq)$ and $I_2(aq)$. Predict the reactions that would you would expect as each is mixed with TBME and solutions of the other halides.

In fact, it would be impossible for your instructor to prepare an aqueous solution of molecular fluorine. Suggest an explanation for this fact.

Name _____ Section _____

Lab Instructor _____ Date _____

EXPERIMENT 28 Nonmetal Reactivities

PRELABORATORY QUESTIONS

1. Look up the color and physical state (solid, liquid, or gas) of the halogens to be tested in this experiment.

2. Fluorine reacts with water to give hydrofluoric acid, HF. What other substance is produced? (*Hint*: It is a gas.)

DATA AND OBSERVATIONS

Part 1

Color of TBME with

    chlorine _____

    bromine _____

    iodine _____

Part 2

Activity tests with chlorine water and TMBE

    chlorine _____

    bromine _____

    iodine _____

Activities with bromine water and TMBE

    chlorine _____

    bromine _____

    iodine _____

Experiment 28 Report Sheet

Activities with iodine water and TBME

    chlorine _____

    bromine _____

    iodine _____

Color produced on addition of starch solution to

    chlorine _____

    bromine _____

    iodine _____

Appearance of solutions after addition of starch solution to

    chloride ion _____

    bromide ion _____

    iodide ion _____

CONCLUSIONS

1. Write your conclusions regarding the relative reactivities of chlorine, bromine, and iodine.

POSTLABORATORY QUESTIONS

1. Write equations for all of the reactions that occurred when chlorine water and bromine water were added to the three halides.

2. What happens when starch solution reacts with the various halides and halogens?

EXPERIMENT 29

Synthesis and Quantitative Analysis of an Iron Compound

Introduction

Despite the theoretical advances that have taken place in recent years, along with increased interest in the subatomic structure of atoms, one of the most exciting areas of chemistry is still synthesizing new compounds and performing analyses to establish their composition. No matter how assiduously we apply our sets of "rules" for nature to follow, the only real answer to the question, "I wonder what will happen if . . ." is to do the experiment. In this activity you will do just that.

In the first part of the experiment, you will prepare separate solutions of two reactants, ferric chloride [iron(III) chloride] and potassium oxalate, then combine the solutions to form the desired product, consisting of potassium, ferric and oxalate ions, and one or more waters of hydration. The newly synthesized compound will be purified by recrystallization.

The purpose of the second and third parts of the experiment is to conduct a series of analyses on the product you synthesized, in order to determine the relative percentages of the constituent parts: oxalate ion, iron(III) ion, and waters of hydration. From this information you can determine the empirical formula of the compound you have prepared. Oxalate will be determined by redox titration, and iron(III) content will be determined spectrophotometrically. The determination of waters of hydration may be directly determined by heating the product for a minimum of 2 hours in a 120° oven to drive off the water, followed by cooling in a desiccator. The remaining constituent, potassium ions, will be determined by difference. When your analyses are complete, you will compare your percentages for the various constituents with the actual values for the expected product of the synthesis.

Water always means distilled or deionized water. When masses are given, try to get as close to the specified amount as you can, but if the directions say "about," then an estimate will do — less is better in such cases.

Procedure Summary

In this experiment you will synthesize an unknown compound of iron, then use a series of analytical techniques to determine the formula of the compound synthesized. The quantities used are extremely small, so great care must be taken throughout the experiment.

Review the discussion in Experiment 20, Analysis by Oxidation-Reduction Titration, for help with Part 2.

234 Experiment 29 Synthesis and Quantitative Analysis of an Iron Compound

Prelaboratory Assignment

Read the Introduction and Procedure sections, and answer the Prelaboratory Questions on the Report Sheet.

Materials

Apparatus

Analytical or milligram balance
Reaction tubes or 100-mm test tubes (2)
Hot sand bath, hot plate, or hot water bath
Boiling sticks
Distilled water (wash bottle)
Microtip pipettes (4)
25-mL volumetric flask
Spectronic 20-D or similar spectrophotometer
Drying oven (110–120°C)
Desiccator
10-mL Erlenmeyer flasks or 20- to 30-mL beakers, as titration vessels (at least 3)
Tubes or cuvettes for use with spectrophotometer

Reagents

Solid iron(III) chloride hexahydrate
Solid potassium oxalate monohydrate
Acetone (wash bottle)
0.2 M sulfuric acid, H_2SO_4
6 M sulfuric acid, H_2SO_4
$KMnO_4$ solution, 2×10^{-5} mol/g
Ferrous ammonium sulfate, $Fe(NH_4)_2(SO_4)_2$
1 M ammonium acetate
1 M hydroxylamine hydrochloride
1,10-phenanthroline

Safety Information

1. **Safety goggles must be worn at all times in the laboratory.**
2. **The sand bath will be hot, though it will not appear so.** The sand has a temperature ≥ 200°C.
3. **Acetone vapors are flammable.** Keep the acetone away from all heat sources.

Procedure

Part 1: Synthesis of the Unknown Compound

1. Dissolve 0.250 ± 0.005 g of solid iron(III) chloride hexahydrate, $FeCl_3 \cdot 6\,H_2O$, in a previously weighed test tube with about 0.5 mL of water. Gentle warming on a sand bath may be necessary to get all of the solid to dissolve. Add more water only as a last resort.

2. Into a separate test tube, place 0.600 g of potassium oxalate monohydrate, $K_2C_2O_4 \cdot H_2O$. Add about 2 mL of water, then using a boiling stick to control the heating, bring the mixture to near boiling to dissolve the solid. Try to maintain the temperature just below the boiling point; if boiling begins, remove the tube and contents from the heat.

3. When all of the oxalate has dissolved, pour the solution quickly and quantitatively into the tube containing the solution of iron(III) chloride. Maintain the combined solutions at near boiling for 1 minute or so, noting any changes in appearance of the mixture. Allow the mixture to cool; crystals should begin to form. If there is no crystallization, try cooling the tube in ice water. If crystals still do not appear, try scratching the walls of the tube with a glass stirring rod.

4. Remove any remaining supernatant liquid by pulling it up into a clean microtip pipette. For best results, expel most of the air from the pipette, then press its tip against the bottom of the tube and allow the liquid to be pulled up slowly into the pipette. Dispose of the liquid as directed by your instructor. Save the pipette for step 6.

5. Set one crystal aside to use as a seed, then recrystallize your product by adding about 1 mL of water to the crystals in the tube. Heat on a sand bath to dissolve the crystals, and allow the crystals to form as before, inducing crystallization by adding the seed crystal to the solution as it cools or by scratching the inside of the tube, as necessary.

6. Once again, remove the liquid from the crystals with the same pipette, disposing of the liquid in the same manner as before. Add a few drops of ice-cold 95% ethanol to your crystals, agitate the tube briefly, then draw off the ethanol with the pipette.

7. Finally, add 5 to 8 drops of ice-cold acetone to the crystals; this will remove the last of the water. Remove the acetone with the pipette and gently warm the tube to drive off the last traces of this volatile solvent. Allow the product to air dry, then weigh the tube and contents to determine the mass of your hydrated crystalline product.

8. Place the tube in a 120°C oven for at least 2 hours, cool it in a desiccator, then weigh it again. Ideally, you should return the tube and product to the oven for an additional hour or more, cool, and reweigh once more. Normally, this is continued until constant mass is achieved, but for the purposes of this experiment, we will dispense with the attempt to achieve constant mass.

Part 2: Determination of Oxalate Ion Content of Product

The oxalate determination and the spectrophotometric determination of iron both require a solution of your product. The same solution may be used for both processes, and is prepared as follows.

9. Dissolve 20 mg (±0.1 mg, if possible) of your anhydrous solid product in a small amount of 0.2 M sulfuric acid in a 25-mL volumetric flask. Once the product is completely dissolved, fill the flask to the mark with 0.2 M sulfuric acid and mix the contents thoroughly by inverting the flask a number of times.

10. To determine the mass percent of oxalate ion in your product, a minimum of three samples must be titrated with a permanganate solution whose concentration is accurately known, in mol MnO_4^-/g of solution. Compare Experiment 20.

11. Label three pipettes, $KMnO_4$ (for permanganate), H_2SO_4 (for 6 M sulfuric acid), and PROD (for your product). With modern top-loading balances it is not difficult to determine the weight of a pipette to ±0.001 g but it is very difficult to measure the volume of a liquid to ± 0.001 mL, so determine the individual masses of the filled $KMnO_4$ and PROD pipettes to the nearest 0.001 g; the mass of the sulfuric acid pipette is not needed.

12. Place between 1.0 and 1.5 g of PROD solution (about one-third of the capacity of the pipette) in a clean, dry 10-mL flask (or small beaker) and add about 0.5 mL (approximately) of 6 M sulfuric acid from your H_2SO_4 pipette. Warm the mixture for 2 to 3 minutes on a hot plate or sand bath to a temperature between 60 and 70°C. Since it is difficult to determine the temperature of such small volumes, discontinue heating when the solution starts to give off steam.

13. While the solution is still hot, begin to add potassium permanganate solution from your $KMnO_4$ pipette, swirling the titration vessel constantly, until you have a pink color that persists for at least 30 seconds. It is important that you keep the contents mixed throughout the titration. If drops of titrant (permanganate) should collect on the sides of the vessel, use a drop or two of H_2SO_4 from your H_2SO_4 pipette to rinse them down. Record the final masses of the $KMnO_4$ and PROD pipettes.

14. Repeat the titration twice more, for a total of three trials. You should use a fresh beaker or flask for each trial, but you will want to rinse each vessel thoroughly immediately after use to prevent formation of permanent stains in the glass.

Part 3: Determination of Iron Content of Product

Iron(III) will form a complex ion with 1,10-o-phenanthroline. This complex absorbs strongly at about 510 nm in the visible range of the spectrum. In this part of the analysis you will form the complex with your product solution, then determine the absorbance of your mixture at 510 nm. By comparing it with a plot of the absorbance of a series of standards, you will determine the relative concentration of iron(II) in your solution. The spectrophotometer usually needs a minimum of 20 minutes to warm up, so it should be turned on early in the laboratory period.

Spectrophotometry The spectrophotometer or colorimeter is used to measure quantitatively how much light a solution absorbs. From this the concentration of light-absorbing ions or molecules in the solution can be calculated. If we define the intensity of light falling on a solution, the incident light, as I_o and the intensity of the light transmitted by the solution as I then, according to the Beer–Lambert law,

$$\log (I_o/I) = \varepsilon\, b\, C = A$$

where ε is the molar absorptivity, a constant that describes the intensity of light absorption at a given wavelength, b is the path length through which the light travels (often 1 cm), and C is the concentration of the absorbing species. The absorbance, A, is the value you will read on the spectrometer. From the above equation, we can see that the absorbance should be a linear function of the concentration if the path length and the molar absorbtivity are constant. In this experiment we will test this assumption. If it is true we say that the absorbing species obeys Beer's law.

Experiment 29 Synthesis and Quantitative Analysis of an Iron Compound   237

Preparation of Solutions   The primary iron standard solution, on which all else relies, must be very carefully and precisely prepared. Because the solution must be very dilute, it is necessary to begin by preparing a stronger solution than is needed; this solution is then diluted to obtain the desired iron concentration.

15. A concentration of 0.025 mg Fe/mL may be attained by dissolving 0.1755 grams of ferrous ammonium sulfate, $Fe(NH_4)_2(SO_4)_2 \cdot 6H_2O$, in a small amount of water in a 1-L volumetric flask, with 2.5 mL of concentrated sulfuric acid. When the solid has dissolved, the flask is filled to the mark with distilled water and the contents mixed thoroughly. (At this point, you have an iron concentration of 0.25 mg/mL.)

16. Using a volumetric pipette, transfer exactly 10.00 mL of this solution to a clean 100-mL volumetric flask and dilute to the mark with water. Again mix the contents thoroughly by inverting the flask a number of times. This solution has the desired 0.025 mg/mL concentration of ferrous ion.

17. You will also need the following solutions; prepare them in the amounts shown:
    1 M ammonium acetate (3.85 g/50 mL)
    1 M hydroxylammonium chloride (3.48 g/50 mL)
    1,10-o-phenanthroline (0.3g/100 mL)

1,10-Phenanthroline

Preparation of Reference Standards and Unknown   As mentioned earlier, the determination of iron content is based on formation of a complex ion of iron(II), yet your compound was the result of a precipitation between iron(III) chloride and potassium oxalate. The function of the hydroxylammonium chloride is to reduce any iron(III) to iron(II); the ammonium acetate is a buffer, which will partially neutralize the acid present in the solution of your product, since the complex of 1,10-phenanthroline only forms with the 2+ oxidation state of iron. The complex absorbs strongly at a wavelength of 510 nm (1 nm = 10^{-9} m).

18. A set of five reference standards will be used to determine the relationship between iron(II) ion concentration and the amount of light absorbed by the solution. The references are to be prepared in 25-mL volumetric flasks; a sixth flask will be used for your unknown. Place one milliliter *each* of the ammonium acetate and hydroxylammonium chloride solutions plus 5 mL of the phenanthroline solution in each flask.

19. To one of the flasks add (volumetric pipette) 1.00 mL of the standard iron solution, to another add 2.00 mL, to a third add 3.00 mL, and so on until you have five reference solutions containing from 1 to 5 mL of the iron standard. Calculate and record the concentrations of iron in each of these flasks in parts per million (ppm). To the sixth flask, add 1.00 mL of your unknown solution remaining from Part A. Fill all six flasks *nearly* to the mark with water, then allow them to stand for a few minutes while you zero the spectrophotometer and set it for zero absorbance (100% transmission) at 510 nm.

20. Fill the reference solution flasks (but not the unknown) to the mark, stopper each flask, and invert them several times to ensure mixing. Determine the absorbance of each of the reference solutions, then make a plot of absorbance versus concentration; the plot should be linear. Now fill your unknown flask to the mark and invert to mix, then determine the absorbance of your unknown. By reference to your plot of absorbance versus concentration, you should be able to arrive at the concentration of iron in your unknown.

21. Use this concentration in parts per million to calculate the mass percent of iron that was in your unknown, assuming the solution that you used has the same density as water. Remember that the mass of the product that you used was originally dissolved in 25 mL of water, and that 1 mL of that solution was used to make 25 mL of the solution that you analyzed spectrophotometrically.

22. Calculate the percentage of iron in $K_3Fe(C_2O_4)_3 \cdot 3\ H_2O$, then calculate the difference between your experimental result from Step 19 and the theoretical percentage for potassium iron(III) oxalate.

Using the Spectronic 20 Spectrophotometer

Turn on the instrument about 20 to 30 minutes before use and set the wavelength to 510 nm. Each time the instrument is used it must be calibrated. With nothing in the sample compartment set the Zero Control to 0%T (transmittance). Use the mirror behind the needle to avoid parallax in the reading. Fill the cuvette with distilled water and place it in the spectrophotometer with the black mark on the cuvette toward yourself. Close the compartment and use the Light Control knob to give a 100%T reading. Remove the cuvette. You are now ready to carry out measurements on your samples.

Using the Spectronic 21 Spectrophotometer

Turn on the instrument about 20 to 30 minutes before use and set the wavelength to 510 nm. Fill the cuvette with distilled water and place it in the spectrophotometer with the black mark on the cuvette toward yourself. It is not necessary to zero this instrument, so close the compartment and adjust the full-scale control (found on the lower left front panel) to give a 100%T reading. Remove the cuvette. You are now ready to carry out measurements on your samples.

Using the Vernier Colorimeter

Connect the Vernier colorimeter to the serial box interface along with the 9-V power supply. The interface box is in turn connected to a serial port on the computer, usually the modem port. Load the computer program **Data Logger** and open the program **Colorimeter** in the *Experimental Files* folder. It will ask you if you want to load the calibration saved with the colorimeter. Select *No*. Under the **Collect** menu select **Calibrate** and then select **Calibrate Now**. Select *Port 1 Only*, wait for the reading to stabilize (±0.001), select **Stable**, type *0*, and select *OK*. Put a cuvette filled with distilled water in the colorimeter with the black mark on the cuvette toward yourself. Close the compartment, set the knob to Blue (470 nm), and repeat the calibration process, but this time type in *100* after the reading stabilizes, followed by *OK*. Your are now ready to read the %T of your samples.

Remove the cuvette, empty it, and rinse it twice with the first iron standard solution. Place the cuvette, after wiping the outside with a tissue, in the colorimeter with the black mark toward you. Close the lid. Click on the **Start** button. When the absorbance value has stabilized click on the Keep button and then enter the concentration in ppm of the solution in the box under **Conc**. Repeat this process for the other four standards and then click the **Stop** button.

Choose **Data A Table** from the main menu and record on your data sheet the absorbance and the concentration of the five solutions. Note whether the curve on the graph goes through the origin. Should it? Choose **Fit** from the **Analyze** menu and under the **Fit** window click **Linear**. Click **Maintain Fit** and note the slope, m, of the least-squares line through your data points.

Now rinse out the cuvette and fill it three-quarters full with your unknown. From the absorbance value found for the unknown and the calibration curve determine the concentration of the unknown. You can do this by dividing the absorbance value by the slope of the least-squares line through your calibration data.

Cleaning Up

Depending on the local laws it may be possible to dispose of these solutions by pouring them down the sink and then flushing with much water. Otherwise collect all solutions in the containers provided.

Name _____ Section _____

Lab Instructor _____ Date _____

EXPERIMENT 29 **Synthesis and Quantitative Analysis of an Iron Compound**

PRELABORATORY QUESTIONS

1. Give directions for preparing 1.00 L of solution that has an iron concentration of 0.250 mg Fe per milliliter, starting with solid ferrous ammonium sulfate, $Fe(NH_4)_2SO_4 \cdot 6H_2O$. Convert your answers to parts per million, ppm, of iron.

2. The oxalate concentration will be determined by redox titration with permanganate in acidic solution. Write the balanced equation for the reaction between oxalate ions and permanganate ions, given that the products are manganese(II) ions and carbon dioxide.

3. Step 7 of the Procedure says to *gently* heat the product to drive off residual acetone. What might happen if the product is heated too strongly?

Experiment 29 Report Sheet

DATA TABLE

Mass of product (from step 8) _____

	Trial 1	Trial 2	Trial 3
Initial mass of $KMnO_4$ pipette	_____	_____	_____
Final mass of $KMnO_4$ pipette	_____	_____	_____
Mass of $KMnO_4$ used	_____	_____	_____
Initial mass of PROD pipette	_____	_____	_____
Final mass of PROD pipette	_____	_____	_____
Mass of PROD used	_____	_____	_____

Standard Solutions

	ppm Fe^{2+}	Absorbance
1	_____	_____
2	_____	_____
3	_____	_____
4	_____	_____
5	_____	_____

Unknown absorbance _____

Concentration of Fe^{2+} in unknown _____

Mass of iron in sample used to make unknown solution. Show calculation.

Mass of iron in product. Show calculations.

Mass of tube and hydrated product (Step 7). _____

Mass of tube and anhydrous product (Step 8). _____

Mass of water removed. _____

Experiment 29   Report Sheet   243

CALCULATIONS AND CONCLUSIONS

1. For each titration, use your data and the balanced equation from Prelaboratory Question 3 to calculate:
   the moles of permanganate ion used

   the moles of oxalate in each sample

   the concentration of oxalate, in mol $C_2O_4^{2-}$/g for your solution.

   Do this for each of your samples, then report the average value as well. If one of your trials differs from the mean of the other two by more than 2% you may report your average based only on the two best trials, but those two must be within 0.5% of each other.

2. Use your average value from Part 2 to calculate the moles and mass of oxalate ion that were in the product sample that you used to make your 25.00-mL solution. From this you can calculate the mass percent of oxalate in your product. (Assume the solution has the same density as water.)

3. Using the mass percentages of iron(III), oxalate, and water, determine the mass percentage of potassium in your product.

4. From the mass percentages of all four constituents, determine the empirical formula of your product. Show calculations.

Experiment 29 Report Sheet

POSTLABORATORY QUESTIONS

1. Calculate the difference between the theoretical value for the mass percentage of oxalate that you calculated in Prelaboratory Question 3 and the value you got in the experiment. Now calculate the percentage difference between the theoretical value and your experimental percentage of oxalate.

2. Oxalate ion is known to act as a reducing agent; it is possible that the initial reaction that formed your bright-green product was an oxidation–reduction process, in which oxalate reduced iron(III) to iron(II), while oxalate was being oxidized to carbon dioxide. Write a balanced equation for the reduction of iron(III) to iron(II) by oxalate ion in acidic solution. Consult a table of standard half-cell potentials to determine whether this reaction is likely to be spontaneous. Could this reaction have taken place?

3. a. Determine the molar excess of oxalate over iron(III) in your synthetic mixture [mol oxalate/mol iron(III)].

   b. Is this excess sufficient to account for reduction of iron *and* for formation of a crystalline iron(II) oxalate monohydrate? Discuss.

4. Calculate the mass percentages of iron, oxalate, and water in potassium iron(III) oxalate trihydrate. (Show calculations.)

EXPERIMENT 30

Partition Coefficient of an Organic Acid

Introduction

Water and ether are not *miscible*. After being shaken together the two will separate to give two layers with the ether on top. Some organic molecules will dissolve in both ether and water. When such a molecule, the *solute*, is dissolved in either solvent and the two layers are shaken together the solute will distribute itself between the two solvents. The purpose of this experiment is to measure quantitatively the *partition coefficient*, which is a measure of the solubilities of the solute in the two solvents.

Inorganic ionic compounds are usually water soluble, but insoluble in nonpolar solvents. Conversely nonpolar organic compounds are not soluble in polar solvents such as water. Some weakly ionic compounds, such as organic acids, are somewhat soluble in both nonpolar and polar solvents.

The nonpolar solvent, *tert*-butyl methyl ether, $(CH_3)_3COCH_3$, TBME, and the polar solvent, water, do not dissolve in each other, but acetic acid, CH_3COOH, will dissolve in both. The object of this experiment is to measure how acetic acid partitions between the two immiscible solvent layers and to study the technique of extraction.

If TBME and water are placed in a test tube they will form two layers, with the TBME on top. If acetic acid is then added to the tube and the mixture shaken, it will be found, after the layers separate, that some of the acid is in the TBME and some in the water. The ratio of the concentration in TBME to that in water is a constant called the *partition coefficient*:

$$k = \frac{\text{Concentration of acid in upper layer (TBME)}}{\text{Concentration of acid in lower layer (water)}}$$

If the top TBME layer is carefully removed and a fresh portion of TBME added and the mixture shaken once more, the acetic acid that was in the water layer should again partition itself between the two layers. In this way the acetic acid can be *extracted* from the water. We will try to predict how much acid is extracted each time knowing what the partition coefficient is and then check this prediction by experiment.

The amount of acetic acid in the water layer will be determined by titration with sodium hydroxide solution according to the following equation:

$$CH_3COOH + Na^+OH^- \rightarrow CH_3COO^-Na^+ + H_2O$$

It is difficult to measure microscale quantities of sodium hydroxide solution to 0.001 mL but it is easy to weigh the solution to 0.001g. In this experiment the solution will be weighed before and after the titration to determine the quantity needed to neutralize the acetic acid.

In the second part of the experiment, because it is not easy to separate layers quantitatively, we will measure the extraction reagents volumetrically, a more rapid process than weighing them.

246   Experiment 30   Partition Coefficient of an Organic Acid

Procedure Summary

A solution of acetic acid is standardized by titration with sodium hydroxide. This acetic acid solution is extracted with an equal weight of TBME and the amount of acid remaining in the aqueous layer is again determined by titration. Knowing how much acid was extracted allows calculation of the distribution coefficient, the ratio of the amount of acid in the ether to that in the water. With this information the effect of multiple extractions can be determined theoretically and by experiment.

Prelaboratory Assignment

Read the Introduction and Procedure sections carefully, and answer the Prelaboratory Question on the Report Sheet.

Materials

Apparatus

Reaction tubes or 10 × 100-mm culture tubes
Corks or stoppers for tubes
Microtip Beral pipettes
Erlenmeyer flasks

Reagents

Acetic acid solution
Standard sodium hydroxide solution
Phenolphthalein

Safety Information

1. **Safety goggles must be worn at all times in the laboratory.**
2. **Handle sodium hydroxide solution with care.** Wash from the skin immediately if there is any contact. Wipe up spills with a damp sponge.
3. ***tert*-Butyl methyl ether is flammable.** Be sure no flames are present when working with it.

Experiment 30   Partition Coefficient of an Organic Acid   247

Procedure

Part 1: Standardization of Acetic Acid Solution

The concentration of acetic acid, in mol/g of solution, is determined by titration with sodium hydroxide solution of known concentration.

1. Fill labeled Beral pipettes with solutions of:
   acetic acid
   standard sodium hydroxide solution (record the concentration in moles of sodium hydroxide per gram of solution, mol/g)
   phenolphthalein indicator solution, 0.1%.

2. Weigh the plastic Beral pipettes containing the acid and base. Place approximately 1 mL of the acetic acid solution in a 10-mL Erlenmeyer flask. This is about one-third the capacity of a pipette. Reweigh the pipette that contains the remaining acid. Add 1 or 2 drops of indicator to the flask and mix well by swirling.

3. Weigh the sodium hydroxide pipette and then titrate the acid solution by adding sodium hydroxide dropwise, swirling the flask after every few drops. The *end point* of the *titration* will be signaled by the appearance of a pink color that persists throughout the body of the solution for at least 30 seconds. Weigh the pipette containing the remaining sodium hydroxide solution.

4. Repeat the titration at least two more times, then calculate the ratio of mass of base solution to mass of acid solution. If the three ratios do not agree within 1%, continue doing titrations until you have three ratios that are within sufficient agreement. If after six trials you are unable to achieve agreement on three trials, consult your instructor.

Part 2: Extractions with TBME

A. Determination of the Partition Coefficient

5. Measure 1 mL of acetic acid solution into a tared (previously weighed) reaction tube or 10 × 100-mm culture tube, weigh the tube and contents, and then add, with a microtip Beral pipette an exactly equal weight of TBME. Stopper the tube with a tightly fitting cork and shake it vigorously for about 2 minutes, then allow the layers to separate. It may be difficult to see the interface between the two layers.

6. With a Beral pipette withdraw most of the lower layer and place it in a tared Erlenmeyer flask. Do not remove any of the top layer material or any of the material at the interface. Weigh the flask plus aqueous layer. Titrate the acetic acid in the flask with standard sodium hydroxide to a phenolphthalein end point.

7. Calculate the concentration of acetic acid in the aqueous layer in mol/g from this titration and the concentration in the TBME layer by the difference between this concentration and the original. The ratio of these two concentrations (concentration of acid in ether/concentration of acid in water) is the partition coefficient, k.

8. Repeat the extraction and titration at least once more. As before, calculate the partition coefficient as directed on page 250. If the values do not agree within five percent, carry out the extraction, titration, and calculation a third time, and then average the results.

B. Extraction with Three Portions of TBME

Knowing the distribution coefficient, calculate the final concentration of acetic acid in the aqueous layer when it is extracted with three equal weights of TBME, then carry out the experiment.

9. To do this mix 1.00 mL (almost exactly 1.00 g) of acetic acid solution and 1.35-mL (1.00 g) of TBME. Shake the mixture well, draw off the lower (aqueous) layer as completely as possible, transfer it to another reaction tube or culture tube. In this fashion extract the aqueous layer twice more with fresh 1.35-mL portions of TBME and then titrate the aqueous layer with standard sodium hydroxide solution to determine if your calculated concentration agrees with experiment.

Cleaning Up

The combined TBME layers should be placed in the organic solvents waste container. The aqueous layers, all of which should be neutral because they have been titrated, can be flushed down the drain.

Name _____ Section _____

Lab Instructor _____ Date _____

EXPERIMENT 30   Partition Coefficient of an Organic Acid

PRELABORATORY QUESTION

1. The density of *tert*-butyl methyl ether is 0.740 g/mL; the density of 1.00M acetic acid is 1.01 g/mL. What volume of *tert*-butyl methyl ether would have the same mass as (a) 1.00 mL of 1.00M acetic acid (b) 1.00 g of 1.00 M acetic acid?

DATA AND OBSERVATIONS

Part 1: Standardization of Acetic Acid Solution

	Trial 1	Trial 2	Trial 3
Initial mass of pipette containing acetic acid solution	_____	_____	_____
Final mass of pipette containing acetic acid solution	_____	_____	_____
Mass of acetic acid solution	_____	_____	_____
Initial mass of pipette containing sodium hydroxide solution	_____	_____	_____
Final mass of pipette containing sodium hydroxide solution	_____	_____	_____
Mass of sodium hydroxide solution	_____	_____	_____
Concentration of sodium hydroxide in mol/g of solution (on label of container)	_____	_____	_____
Concentration of acetic acid solution in mol/g of water (show calculation)	_____	_____	_____

Experiment 30   Report Sheet

Part 2: Extractions with TBME

A. Determination of the Partition Coefficient

	Trial 1	Trial 2
Mass of water layer after extraction	_____	_____
Initial mass of pipette containing sodium hydroxide solution	_____	_____
Final mass of pipette containing sodium hydroxide solution	_____	_____
Mass of sodium hydroxide solution	_____	_____
Concentration (mol/g) of acetic acid in water layer after extraction (show calculation)	_____	_____
Concentration (mol/g) of acetic acid in TBME layer (show calculation)	_____	_____
Partition coefficient, k (show calculation)	_____	_____

B. Extraction with Three Portions of TBME

Predicted concentration of acetic acid in aqueous layer after three extractions with TBME (show all calculations)

Mass of aqueous layer after three extractions _____

Initial mass of pipette containing sodium hydroxide solution _____

Final mass of pipette containing sodium hydroxide solution _____

Mass of sodium hydroxide solution _____

Concentration of acetic acid in water layer after three extractions (show calculation) _____

What would the concentration of acetic acid in the water have been if the original aqueous solution of acetic acid were extracted with one 4.0-mL portion of TBME rather than twice with 2-mL portions? (Show calculations.)

CONCLUSIONS

1. An organic acid is to be isolated from an aqueous solution by extraction with TBME. The distribution coefficient for this acid, k, is 1.5. How many moles of the acid would be extracted from 3 mL of an aqueous solution containing 10 mmol of acid per gram of solution if the extraction were conducted using (a) three 1-mL portions of TBME and (b) one 3-mL portion of TBME? *Suggestion*: For each extraction, let x equal the mmol extracted in the TBME layer. In case (a), the concentration in the ether layer is x mmol/g and that in the water layer is $(10-x)$ mmol/g; the ratio of these quantities is equal to the distribution coefficient, k.

2. How much TBME would be required to remove 98.5% of the acid from the water in Question 1 using just one extraction?

EXPERIMENT 31

Analysis of Commercial Vinegar

Introduction

Vinegar is an aqueous solution of acetic acid, often produced by the fermentation of alcohol (wine). Federal regulations stipulate that commercial vinegar contain at least 5% acetic acid by mass. To ensure the law is being complied with commercial vinegar is titrated with base in order to calculate the acetic acid concentration.

Acid-base titrations are one of the fundamental skills a chemist learns. When a strong acid, such as hydrochloric acid, is allowed to react with a strong base, such as sodium hydroxide, NaOH, the products are water and a salt formed from the cation of the base and the anion (conjugate base) of the acid. In the example cited here, the salt would be sodium chloride, NaCl. When the amount of acid and base that react with each other are exactly equal the solution will be near neutral. This could be determined with a pH meter or, more conveniently, an indicator such as phenolphthalein that will change color at the equivalence point.

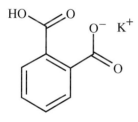

Potassium hydrogen phthalate

It is effectively impossible to prepare solutions of most solutes in accurately-known concentrations. For example, sodium hydroxide tends to absorb both water and carbon dioxide from the air. The former affects the mass of material, but does not change it chemically; the latter reacts with the NaOH, forming both bicarbonate and carbonate ions. For that reason, sodium hydroxide is prepared in approximately the concentration desired, then standardized (titrated) against a "primary standard." A primary standard is a substance that is obtainable in high purity and whose mass can be measured with high precision; the one usually chosen for standardizations of basic solutions is potassium hydrogen phthalate, also called potassium biphthalate, potassium acid phthalate, or simply KHP. The molecular formula is $KC_8H_5O_4$.

For cases in which very small amounts of solutions are being used, volumes are difficult to measure with good precision, hence molarity is not a convenient unit for solution concentration. Instead, we will measure concentrations in terms of moles of solute per gram of solution. It will be a simple matter, then, to determine the moles of titrant used, by measuring the mass of solution delivered.

Procedure Summary

This experiment has two parts: the first involves standardization of an unknown sodium hydroxide solution ("BASE"), and is truly experimental in that part of your prelaboratory assignment is to calculate the concentrations of solutions needed for the experiment to work. In Part II you will use the same sodium hydroxide solution to determine the strength of commercial vinegar.

Prelaboratory Assignment

Read the Introduction and Procedure; answer Prelaboratory Questions 1-5 on the Report Sheet before you come to the laboratory.

Materials

Apparatus

Milligram balance
50-mL beaker or flask
Microtip pipettes (2)
10-mL Erlenmeyer flasks or small beakers as titration vessels
25-mL volumetric flask

Reagents

Potassium acid phthalate
Distilled water (wash bottle)
NaOH (to be standardized)
Phenolphthalein solution, in microtip pipette
Commercial white vinegar

Safety Information Sodium hydroxide is extremely caustic; all contact with skin and clothing is to be avoided. In case of accidental contact, wash the affected area with liberal amounts of water and notify your instructor.

Procedure

Part I: Standardization of Sodium Hydroxide by Potassium Hydrogen Phthalate

Preparation of Standard (May be done by instructor.)

1. Accurately weigh the potassium hydrogen phthalate primary standard (which you calculated in Pre-lab question 2) needed to make 25.0 mL of 0.100 M solution. Use KHP that

has been dried at 110°C for several hours then allowed to cool. This removes water that may have been absorbed from the air, and will increase the precision of the standardization. Place the solid in a previously-weighed (tared), clean, dry 50-mL beaker or flask, and add water to make approximately 25.0 mL total volume. Swirl to dissolve, then reweigh the beaker and solution. Calculate and record the concentration of the solution in moles of KHP per gram of solution.

Standardization of NaOH.

2. Label separate Beral pipettes for KHP and BASE, and fill them respectively with the standard potassium hydrogen phthalate, and the sodium hydroxide solution whose concentration is to be determined. Wipe the tips of the pipettes dry, then determine their individual masses. If the KHP standard was prepared for you, be sure to record its concentration in moles of potassium hydrogen phthalate per gram of KHP solution.

3. Deliver 1.0-1.5 grams of the standard KHP solution to each of three 10-mL flasks or other suitable titration vessels (such as 20- or 30-mL beakers). Record the mass of solution placed in each vessel as determined by subtracting the initial and final masses of pipette and contents. You may have to refill the pipette between samples.

4. Add one drop of phenolphtalein indicator to one of your samples, then begin adding sodium hydroxide solution, a few drops at a time, swirling the flask after each addition. Continue in this fashion until the appearance of a faint magenta color which persists throughout the solution for at least 30 seconds. This indicates that the solution has reached the equivalence point and that the number of moles of base added exactly equals the number of moles of acid present. Record the final mass of your BASE pipette and contents. Repeat the titration with each of your other two KHP samples, refilling the BASE pipette between trials as necessary. Remember to reweigh the pipette after each filling.

5. Compare the relative masses of the two solutions used in each trial. If the ratios of mass of KHP solution to mass of NaOH solution agree to ±1%, proceed with the next part of the Procedure; if they do not, carry out additional titrations until you have at least three within satisfactory agreement.

Part II: Determination of the Mass Percent of Acetic Acid In Commercial Vinegar

6. Determine the mass of a clean, dry 25-mL volumetric flask, then use a clean volumetric or graduated (Mohr) pipette, to transfer exactly 2.50 mL of a commercial vinegar product to the flask. Use distilled water to bring the volume of solution in the flask up to exactly 25.00 mL, then invert the stoppered flask several times to ensure complete mixing. Determine the mass of flask and contents.

7. Following the procedure of Part I. prepare three fresh 10-mL flasks, each containing between 1.0 and 1.5 grams of the diluted vinegar. Add phenolphthalein to each and titrate with the sodium hydroxide which you standardized in Part I. As before, check the ratios of acid to base for the three to be sure that they are in good agreement, and no further trials are needed. If agreement is not satisfactory, do additional titrations as necessary.

Cleaning Up

At the conclusion of each part of the procedure, the contents of your titration vessels are safe enough to be rinsed down the drain with water. The remaining sodium hydroxide solution in your beaker, however, is not, and must be neutralized. This is easily accomplished by mixing it with the same vinegar solution that was used in the experiment. One drop of phenolphthalein added to the solution of sodium hydroxide in the beaker will tell you when neutralization is complete. Rinse all equipment thoroughly with water, then leave it on paper towels to drain for the next class.

Name _____   Section _____

Lab Instructor _____   Date _____

EXPERIMENT 31   Analysis of Commercial Vinegar

PRELABORATORY QUESTIONS

1. Copy the structural formula for potassium hydrogen phthalate and identify the lone acidic hydrogen on the structure.

2. Calculate the mass of KHP needed to make 25.0 mL of 0.100 M solution (watch significant figures).

3. Assuming a density of 1.00 g mL^{-1} for both solutions, what mass of 0.100 M NaOH solution will be needed to exactly neutralize 1.00 gram of the KHP solution?

4. Use the molarity and density values from questions 2 and 3 to calculate the soluton concentrations for NaOH and KHP in moles of solute per gram of solution. Assume no dilution effects.

5. Commercial vinegar is approximately 5% acetic acid by mass.

   a. Assuming a density of 1.0 g mL^{-1}, calculate the concentration of acetic acid in vinegar, in mol L^{-1}.

   b. If 1.0 mL of an acetic acid solution of this concentration is to be neutralized by 0.10 M NaOH, what volume of the base will be required?

257

258   Experiment 31   Report Sheet

c. A Beral-type pipette holds about 3.5-4.0 grams of solution; will one pipette-full of base be enough to complete the titration of full-strength vinegar?

OBSERVATIONS AND DATA

Part I:

	Trial 1	Trial 2	Trial 3
a. Mass of KHP	_____	_____	_____
Mass of beaker	_____	_____	_____
Mass of beaker and solution	_____	_____	_____
Concentration of KHP solution	_____	_____	_____

b. **Titration Data**

	Trial 1	Trial 2	Trial 3
Mass of KHP pipette before titration	_____	_____	_____
Mass of KHP pipette after titration	_____	_____	_____
Mass of KHP solution used	_____	_____	_____
Mass of NaOH pipette before titration	_____	_____	_____
Mass of NaOH pipette after titration	_____	_____	_____
Mass of NaOH solution used	_____	_____	_____
Ratio: Mass of KHP/Mass of NaOH	_____	_____	_____

[Note: If additional trials are run, attach a separate data sheet.]

Part II: Titration Data

	Trial 1	Trial 2	Trial 3
Mass of Vin pipette before titration	_____	_____	_____
Mass of Vin pipette after titration	_____	_____	_____
Mass of Vin solution used	_____	_____	_____
Mass of NaOH pipette before titration	_____	_____	_____
Mass of NaOH pipette after titration	_____	_____	_____
Mass of NaOH solution used	_____	_____	_____
Ratio: Mass of Vin/Mass of NaOH	_____	_____	_____
Mass of Vin pipette before titration	_____	_____	_____
Mass of Vin pipette after titration	_____	_____	_____
Mass of Vin solution used	_____	_____	_____
Mass of NaOH pipette after titration	_____	_____	_____
Mass of NaOH solution used	_____	_____	_____
Ratio: Mass of Vin/Mass of NaOH	_____	_____	_____
Mass of Vin pipette before titration	_____	_____	_____
Mass of Vin pipette after titration	_____	_____	_____
Mass of Vin solution used	_____	_____	_____
Mass of NaOH pipette before titration	_____	_____	_____
Mass of NaOH pipette after titration	_____	_____	_____
Mass of NaOH solution used	_____	_____	_____
Ratio: Mass of Vin/Mass of NaOH	_____	_____	_____

[Note: If additional trials are run, attach a separate data sheet.]

Experiment 31 Report Sheet

CALCULATIONS AND CONCLUSIONS

1. Calculate the concentration of the sodium hydroxide solution in moles of NaOH per gram of BASE solution. Carry out the calculation for each of the three trials used to get the final value. In addition, calculate the average deviation for the three trials and report the concentration of the sodium hydroxide including average deviation.

_____ ± _____ M

2. Use your data from Part II to determine the concentration of the diluted vinegar in moles of acetic acid per gram of solution. Use the result to determine the mass percent of acetic acid in the original (undiluted) vinegar. Report the individual values and the average percentage, including average deviation.

_____ ± _____ %

ANALYSIS AND RESULTS

1. The Introduction describes two successive side-reactions that can change the make-up of sodium hydroxide.

    a. Write balanced equations showing: (i) the reaction between carbon dioxide and hydroxide ions to form bicarbonate ions; and (ii) the further reaction between bicarbonate and excess hydroxide to form carbonate ions and water.

    b. What effect will these have on the concentration of a solution made by dissolving 0.400 grams of solid sodium hydroxide in water to make 100 mL of solution? What approximate molarity would you expect such a solution to have?

2. The molar solubility of carbon dioxide in water is about 0.033 mol L^{-1} at room temperature. Would you expect the solubility to be higher or lower in a solution of sodium hydroxide? Explain.

3. In preparing a sodium hydroxide solution that is to be used for quantitative work, the usual technique is to boil the water first.

    a. Suggest a reason why boiled water might be preferable to water that has been allowed to stand in contact with the air.

    b. Consider the equations you wrote in response to Question 1: what effect, if any, would the presence of carbonates and bicarbonates have on the ability of the solution to neutralize hydronium ions? Explain.

4. Explain, using appropriate equations, why a white crust will sometimes form when sodium hydroxide which has been exposed to air is dissolved in 'hard' tap water (water containing substantial amounts of calcium ions and/or magnesium ions).

5. How would your experimental procedure have been affected if you had used the vinegar directly from the bottle, without dilution? Explain.

6. Identify the conjugate acid-base pairs in each of the following: (a) the reaction between hydroxide ions and hydrogen phthalate ions, which may be represented as $HC_8H_4O_4^-$, with the first H being the acidic hydrogen; (b) the reaction between (molecular) acetic acid and hydroxide ions.

EXPERIMENT 32

Gas Chemistry: Carbon Dioxide

Introduction

In this experiment carbon dioxide gas will be generated and its properties investigated. Look up the properties of carbon dioxide in a chemistry reference book and fill in the "Properties of Carbon Dioxide" questions on the Report Sheet.

Carbon dioxide is one of the top 25 chemicals produced by industry. One source is fermentation. For each molecule of glucose fermented, two molecules of ethanol and two of carbon dioxide are produced. The carbon dioxide can be captured and condensed, under almost 1,000 pounds per square inch pressure, to a liquid. This material is then used to fill fire extinguishers, in carbonated beverages, and for a multitude of other uses.

The atmosphere of the earth is about 0.033% carbon dioxide by volume. Although this is not much, it furnishes the carbon source for photosynthesis and it serves a very useful function in the earth's atmosphere. The earth would be very cold indeed if it were not for carbon dioxide. Sunlight penetrates the earth's atmosphere and heats the planet. But carbon dioxide prevents some of this heat from being reradiated to space, trapping the sun's heat in the atmosphere. In recent decades humans may have upset the delicate balance between heat absorbed and heat radiated by increasing the amount of carbon dioxide ever so slightly through the burning of fossil fuels. An increase in the carbon dioxide concentration means more heat trapped and thus an increase in the temperature of the earth. This "greenhouse effect," if it does indeed exist, may have serious long-term effects on the future of the human race.

One hundred milliliters of water will dissolve about 90 mL of carbon dioxide gas at room temperature. The resulting solution is acidic by virtue of the ionization of carbonic acid, a weak acid.

$$CO_2 + H_2O \rightarrow H_2CO_3$$
$$H_2CO_3 \rightleftharpoons H^+ + HCO_3^-$$
$$HCO_3^- + H_2O \rightleftharpoons CO_3^{2-} + H_3O^+$$

At low temperature and/or under pressure carbon dioxide is more soluble in water. This is the basis for carbonated beverages. When the pressure is released by removing the cap on a beverage the carbon dioxide solution becomes supersaturated with respect to the gas, which comes out of solution as bubbles that form on any solid particles or sharp points such as scratches in the container.

A number of minerals are carbonates. Calcite is crystalline calcium carbonate. Limestone is the more common form of the same compound. Heating calcium carbonate to red heat causes it to lose carbon dioxide to form calcium oxide, which is called lime or quicklime:

$$CaCO_3(s) \rightarrow CO_2(g) + CaO(s)$$

This lime is very reactive toward water and in an exothermic reaction forms calcium hydroxide or slaked lime:

$$CaO(s) + H_2O(l) \rightarrow Ca(OH)_2(s)$$

Lime is used in plaster, mortar, and cement. It reacts with water and "sets" to the hydroxide, which, over a long period of time, reacts with carbon dioxide in the air to form the rock-hard carbonate:

$$Ca(OH)_2(s) + CO_2(g) \rightarrow CaCO_3(s) + H_2O(l)$$

Carbon dioxide also reacts in a similar way with barium hydroxide to form white insoluble barium carbonate. The formation of this white precipitate from a saturated solution of the hydroxide is a common test for carbon dioxide:

$$Ba(OH)_2(aq) + CO_2(g) \rightarrow BaCO_3(s) + H_2O(l)$$

Calcium carbonate (limestone) will dissolve in carbonic acid solution to form soluble calcium bicarbonate. When the carbon dioxide leaves, due to heat or lower pressure, the calcium carbonate comes out of solution. This accounts for the deposit of calcium carbonate in hot water pipes and tea kettles and the formation of stalagtites and stalagmites in caves.

$$CaCO_3(s) + H_2CO_3(aq) \rightleftarrows Ca(HCO_3)_2(aq)$$

$$H_2CO_3(aq) \xrightarrow{\text{heat or low pressure}} H_2O(l) + CO_2(g)$$

Procedure Summary

Carbon dioxide is generated by the action of acid on a carbonate. It is then allowed to react with a pH indicator and barium and calcium ions.

Prelaboratory Assignment

Read the Introduction and Procedure sections carefully, answer the Prelaboratory Question on the Report Sheet, and fill out the Properties of Carbon Dioxide section on the Report Sheet.

Materials

Apparatus

Gas generation and reaction apparatus (see Figure 32.1).

Reagents

Bromthymol Blue indicator
Calcium carbonate
3 M hydrochloric acid
Saturated calcium hydroxide solution
Saturated barium hydroxide solution

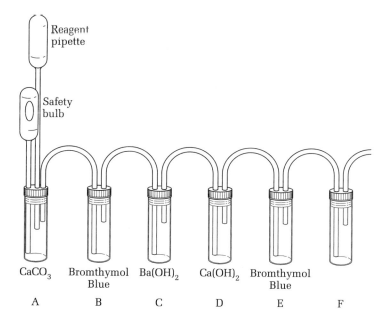

Figure 32.1 Gas generation and reaction apparatus, carbon dioxide.

Safety Informaton

1. **Safety goggles must be worn at all times in the laboratory.**
2. **Handle hydrochloric acid with care; it is very corrosive.** Wipe up spills immediately with a damp sponge.

Experiment 32 Gas Chemistry: Carbon Dioxide

Procedure

Carbon dioxide will be generated by the reaction of calcium carbonate with hydrochloric acid:

$$CaCO_3(s) + 2\ HCl(aq) \rightarrow CaCl_2(aq) + H_2O(l) + CO_2(g)$$

1. Assemble the apparatus shown in Figure 32.1. Note that in vials C and D a long tube that extends nearly to the bottom of the vial is employed. In vial A place about 0.2 g of calcium carbonate. While the apparatus is apart, draw about 1 mL of 3 M hydrochloric acid into the reagent pipette. In vial B place a few drops of deionized water and 2 drops of Bromthymol Blue indicator solution, which is used to measure pH. Bromthymol Blue solution is yellow below pH 6.0, green at pH 7, and blue at pH 7.5 and above. In vial C place about 1 mL of saturated barium hydroxide solution, in vial D place about 1 mL of saturated calcium hydroxide solution (limewater), and in vial E place a few drops more deionized water and 2 more drops of Bromthymol Blue indicator solution.

2. Slowly add the 3 M HCl to the calcium carbonate from the reagent pipette. Note whether there is a reaction in vial A and note the sequence of changes in vials B, C, D and E.

Cleaning Up

Empty the vials into the waste container in the hood and then rinse out all vials, tubes, and pipettes with water.

Name _____ Section _____

Lab Instructor _____ Date _____

EXPERIMENT 32   Gas Chemistry: Carbon Dioxide

PRELABORATORY QUESTION

1. If the tube leading from vial B to C should become blocked while carbon dioxide is being generated in vial A, would carbon dioxide create a high pressure and blow apart the apparatus? Explain your answer.

PROPERTIES OF CARBON DIOXIDE

Molecular weight _____ g/mol

Appearance _____

Density of gas _____ g/L

Melting point of solid _____

Solubility in water _____ g/100 mL water at 25°C

OBSERVATIONS

What change occurred in vial A when the HCl was added?

What change occurred in vial B?

What change occurred in vial C?

What change occurred in vial D?

What conclusion can you draw about the pH of an aqueous solution that is exposed to carbon dioxide?

What conclusion can you draw about a saturated solution of barium hydroxide that is exposed to carbon dioxide?

CALCULATIONS AND CONCLUSIONS

1. How many moles of hydrochloric acid are contained in 1 mL of 3 M hydrochloric acid?

2. How many moles of calcium carbonate will react with 3×10^{-3} moles (0.003 moles or 3 millimoles) of hydrochloric acid?

3. From the equation for the reaction, what weight of calcium carbonate will react with 1 mL of 3 M hydrochloric acid?

4. How many moles of carbon dioxide are produced in this reaction?

5. At standard temperature and pressure, what is the volume of carbon dioxide produced in this reaction?

6. Dry ice is a solid that sublimes to a gas without passing through the liquid state. What are the conditions needed to observe carbon dioxide as a liquid?

7. Carbon dioxide is a vital part of the life process. If you do not know already, guess the mole percent composition of carbon dioxide in air and the mole percent of argon in the air. You may be surprised at the answer which you can look up in your textbook.

8. When limestone (calcium carbonate) is heated to very high temperature it forms lime (calcium oxide). When calcium oxide and sand are mixed with water mortar is formed and the calcium oxide is converted to calcium hydroxide. Over a very long period of time the calcium hydroxide is converted to another substance that is as hard as a rock. What is this substance? Write an equation for its formation.

EXPERIMENT 33

Gas Chemistry: Ammonia

Introduction

In this experiment ammonia gas will be generated and its properties and reactions investigated.

Ammonia is a pungent gas that can be condensed to a liquid at –33°C (–27°F). In the United States it is the fifth most common industrial chemical being produced in quantities exceeding 15 million tons per year. Most of it is used as fertilizer, either as the gas itself or in combination with nitric acid as ammonium nitrate. Ammonium nitrate is one-third nitrogen, by weight, all of it in a biologically available form. Ammonium nitrate is also an explosive, technically a blasting agent, which when mixed with 6% fuel oil and set off with a booster, has about 40% of the power of dynamite, but is much less expensive. A ship full of this material exploded in Texas City, Texas, in 1947, wiping out the entire town, and in 1995 a truck full was responsible for the deaths of nearly 170 people in Oklahoma City.

Its high heat of vaporization, stability, low density, and low corrosiveness make ammonia a valuable refrigerant and it is employed in large quantity to make sodium carbonate in the Solvay process.

Ammonia is made by the Haber process in which nitrogen gas from the air is reacted with hydrogen at 400°C and 250 atmospheres pressure in the presence of a catalyst consisting of iron, aluminum, and potassium oxides. The hydrogen is made by treating methane with steam at high temperatures.

$$CH_4(g) + 2\ H_2O(g) \rightarrow 4\ H_2(g) + CO_2(g)$$

$$N_2(g) + 3\ H_2(g) \rightarrow 2\ NH_3(g)$$

The formation of ammonia is an exothermic reaction but much energy must be expended in this process of nitrogen fixation. Breaking the N≡N bond requires 941 kJ/mol and breaking the H–H bond requires 432 kJ/mol. On the other hand nitrogen-fixing bacteria in peas, beans, and alfalfa do it at normal, outdoor temperature and atmospheric pressure. Trying to emulate this biochemical process in the laboratory has been a long-standing goal of chemical research.

In the present experiment ammonia will be made by the reaction of calcium hydroxide with ammonium chloride:

$$2\ NH_4Cl(aq) + Ca(OH)_2(aq) \rightarrow 2\ NH_3(g) + CaCl_2(aq) + 2\ H_2O(l)$$

Ammonia has a trigonal pyramidal structure because the nitrogen atom is sp^3 hybridized. The H–N–H angle is 107°, not far from the 109.5° tetrahedral angle with the three hydrogen atoms occupying three apices of the tetrahedron, while a lone pair of electrons occupies the fourth. This lone pair of electrons makes ammonia a Lewis base and can be shared with a proton to form the ammonium ion:

$$:NH_3 + H^+ \rightarrow NH_4^+$$

The electron pair also can be donated to a variety of metal ions to form coordination compounds. In these cases the ammonia molecules are called *ligands*.

$$Co^{3+} + 3Cl^- + 6(:NH_3) \rightarrow [Co(:NH_3)_6]^{3+} + 3Cl^-$$

Many coordination compounds of this type have an octahedral structure.

Ammonia and chloride ion are described as *monodentate ligands* (single-toothed) in compounds of this type. Other ligands such as $H_2NCH_2CH_2NH_2$ are *bidentate* because the two nitrogens can "bite" the cobalt at two places.

The ligands can distribute themselves about the cobalt in various ways giving rise to stereoisomers, molecules that differ from one another by the arrangement of the various ligands in space. When the chlorine atoms are *trans* to each other this particular complex is green, when *cis* to each other it is violet.

trans-Isomer *cis*-Isomer

$[Co(NH_3)_4Cl_2]Cl$, tetraamminedichlorocobalt(III) chloride

Procedure Summary

Ammonia is generated by the reaction of calcium hydroxide with ammonium chloride. The gas is allowed to react with a pH indicator and then cobalt, nickel, and copper ions. It is neutralized with hydrochloric acid.

Prelaboratory Assignment

Read the Introduction and Procedure sections carefully and answer Prelaboratory Questions on the Report Sheet.

Materials

Apparatus

Gas generation and reaction apparatus (see Figure 33.1)

Reagents

Ammonium chloride
Calcium hydroxide
Universal indicator
1 M $CoCl_2$ solution
1 M $NiCl_2$ solution
1 M $CuSO_4$ solution
Concentrated hydrochloric acid

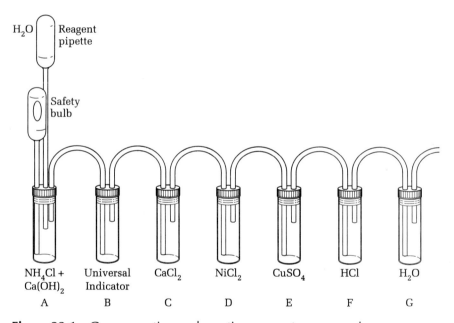

Figure 33.1 Gas generation and reaction apparatus, ammonia.

Safety Information
1. Safety goggles must be worn at all times in the laboratory.
2. **Handle hydrochloric acid with care; it is very corrosive.** Wipe up spills immediately with a damp sponge.

Procedure

1. In vial A (see Figure 33.1) place about 1 g of ammonium chloride and 1 g of calcium hydroxide. In vial B place about 3 drops of water and 1 drop of Universal Indicator. In vial C place 0.3 mL of a 1 M $CoCl_2$ solution, in vial D place 0.3 mL of a 1 M $NiCl_2$ solution, in vial E place 0.3 mL of a 1 M $CuSO_4$ solution, in vial F place a few drops of concentrated hydrochloric acid, and then half fill vial G with water.

2. To start the reaction add about a milliliter of water from the reagent pipette in Vial A and make sure the bottom of the safety bulb is covered so ammonia will not escape from the apparatus. To increase the rate of ammonia generation, heat vial A in a small beaker of warm water. Stop the heating when you see evidence of a reaction in vial E. Cool vial A and add water to it through the opening in the safety bulb to half fill vial A.

Cleaning Up

Take the entire apparatus to the hood, open up each vial, and empty the contents of each one into the waste container provided. Rinse each vial and the connecting tubes with water.

Name _____ Section _____

Lab Instructor _____ Date _____

EXPERIMENT 33 Gas Chemistry: Ammonia

PRELABORATORY QUESTIONS

1. If the tube leading from vial A to B should be come blocked while ammonia is being generated in vial A, would the ammonia create a high pressure and blow apart the apparatus?

2. Ammonia is very soluble in water. How does the procedure insure that ammonia will be transferred from Vial A to Vials B–F?

PROPERTIES OF AMMONIA

Molecular weight _____ g/mol

Appearance _____

Density of gas _____ g/L

Boiling point _____

Solubility in water _____ g/100 mL water at 25°C

Experiment 33   Report Sheet

OBSERVATIONS

1. Odor of ammonium chloride _____

   Odor of calcium hydroxide _____

   Odor of the mixture after it was placed in vial A _____

   Conclusion

2. Appearance of the mixture in vial A at the beginning and end of the reaction _____

   Conclusion

3. Appearance of the Universal Indicator in vial B at the beginning and at the end of the reaction _____

   Conclusion

4. From the chart supplied with the Universal Indicator, what is the approximate pH of the solution in vial B at the beginning and at the end of the reaction? _____

   Conclusion

5. What is the color of the solution in vial C at the beginning of the reaction? _____

   After it has reacted with ammonia? _____

   Conclusion

Experiment 33   Report Sheet   277

6. What is the color of the solution in vial D at the beginning of the reaction?   _____

   After it has reacted with ammonia?   _____

   Conclusion

7. What is the color of the solution in vial E at the beginning of the reaction?   _____

   After it has reacted with ammonia?   _____

   Conclusion

8. What is the appearance of the concentrated hydrochloric acid in vial F at the beginning of the reaction?

   _____

   At the end of the reaction?   _____

   Conclusion

POSTLABORATORY QUESTIONS

1. Write an equation for the reaction of ammonia with water that might explain the pH observed in vial B.

278   Experiment 33   Gas Chemistry: Ammonia

2. Can you propose a structure for the coordination compound that might have formed in each of the following?

   a. vial C

   b. vial D

   c. vial E

3. Write an equation for the reaction that has occurred in vial F.

EXPERIMENT 34

Gas Chemistry: Sulfur Dioxide

Introduction

In this experiment sulfur dioxide gas will be generated and some of its properties and reactions studied. Look up the physical properties of the reagents used in this experiment in a chemistry reference book and fill in the "Physical Properties of Reagents" questions on the Report Sheet.

Coal contains sulfur and when burned this sulfur is converted to sulfur dioxide. The sulfur dioxide dissolves in the water in clouds to form sulfurous acid, which is then oxidized by oxygen to sulfuric acid, which will eventually fall as rain. This acid rain produced by the burning of fossil fuels is a major problem throughout the industrialized world. Acid rain can kill fish and evergreens and even dissolve the marble statues of ancient Greece. It can be removed from the stack gases of power plants by reaction with limestone (calcium carbonate) to form calcium sulfate (gypsum). It can also be oxidized to sulfur trioxide, SO_3, which, when dissolved in water, gives sulfuric acid:

$$SO_3(s) + H_2O(l) \rightarrow H_2SO_4(l)$$

Sulfur dioxide will be generated in this experiment not by the combustion of sulfur but by the action of acid on a sulfite.

$$2\ HCl(aq) + Na_2SO_3(aq) \rightarrow 2\ NaCl(aq) + H_2O(l) + SO_2(g)$$

Sulfur dioxide will react with water to form sulfurous acid:

$$SO_2(g) + H_2O(l) \rightarrow H_2SO_3(aq)$$

This acid dissociates in two steps to give hydrogen ions. The pK of the first step indicates that sulfurous acid is an acid intermediate in strength between hydrochloric and acetic.

$$H_2SO_3(aq) + H_2O(l) \rightleftarrows H_3O^+(aq) + HSO_3^-(aq) \quad pK_1 = 1.76$$
$$HSO_3^-(aq) + H_2O(l) \rightleftarrows H_3O^+(aq) + SO_3^{2-}(aq) \quad pK_2 = 7.20$$

Sulfur dioxide can function as a reducing agent in a number of reactions; in the process it is converted to sulfate ion. Before the advent of modern bleaches, "brimstone" (sulfur) was burned in the presence of such items as straw and palm leaf to bleach them before they were braided to make hats.

Sulfur dioxide will reduce the dark purple permanganate ion. In the process, sulfur dioxide is converted to sulfate ion, SO_4^{2-}. Depending on the pH of the system, the manganese in MnO_4^- may be reduced to the 2+, 4+ or 6+ oxidation state.

In acidic solution the following reaction takes place:

$$2\,MnO_4^-(aq) + 5\,SO_2(g) + 2H_2O(l) \rightarrow 2\,Mn^{2+}(aq) + 5\,SO_4^{2-}(aq) + 4H^+(aq)$$

In neutral solution it is the following:

$$2\,MnO_4^-(aq) + 3\,HSO_3^-(aq) + OH^-(aq) \rightarrow 2\,MnO_2(s) + 3\,SO_4^{2-}(aq) + 2\,H_2O(l)$$

And in basic solution it is this reaction:

$$2\,MnO_4^-(aq) + 2\,OH^-(aq) + SO_3^{2-}(aq) \rightarrow 2\,MnO_4^{2-}(aq) + SO_4^{2-}(aq) + H_2O(l)$$

Similarly dichromate is reduced by sulfur dioxide to chromic ion:

$$Cr_2O_7^{2-}(aq) + SO_2(g) + H^+(aq) \rightarrow Cr^{3+}(aq) + HSO_4^-(aq) + H_2O(l)$$
Unbalanced equation

As an exercise we leave it to you to balance this last equation (Prelaboratory Question).
Other colored metal ions can also be reduced to a variety of oxidation states. An interesting project might be to prepare acidic, neutral, and basic solutions of vanadanate ion (using ammonium vanadanate) and try to deduce exactly what is occuring.
Another project might be to investigate the reaction of an aqueous solution of iodine with sulfur dioxide. Try the reaction and then see if you can propose an equation that would satisfy your observations.

Procedure Summary

Sulfur dioxide is generated by the action of acid on a sulfite. It is then allowed to react with a pH indicator and permanganate and dichromate ions at various pHs.

Prelaboratory Assignment

Read the Introduction and Procedure sections carefully, answer the Prelaboratory Questions son the Report Sheet, and fill out the Physical Properties of Reagents section on the Report Sheet.

Materials

Apparatus

Gas generation and reaction apparatus (see Figure 34.1)
1-mL calibrated pipette, glass or Beral

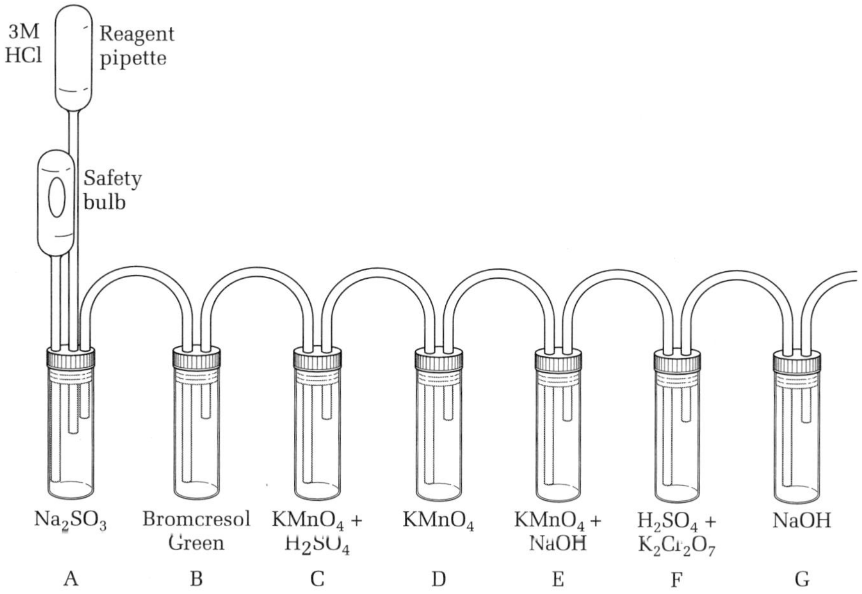

Experiment 34 Gas Chemistry: Sulfur Dioxide

Figure 34.1 Gas generation and reaction apparatus, sulfur dioxide.

Vial labels (left to right):
- A: Na_2SO_3
- B: Bromcresol Green
- C: $KMnO_4$ + H_2SO_4
- D: $KMnO_4$
- E: $KMnO_4$ + NaOH
- F: H_2SO_4 + $K_2Cr_2O_7$
- G: NaOH

Pipette: 3M HCl Reagent pipette, Safety bulb

Reagents

Sodium sulfite
Bromcresol Green indicator
0.01 M potassium permanganate solution
3 M sulfuric acid
2 M sodium hydroxide solution
Potassium dichromate (optional)
3 M hydrochloric acid

Safety Information

1. **Safety goggles must be worn at all times in the laboratory.**
2. **Handle hydrochloric acid with care; it is very corrosive.** Wipe up spills immediately with a damp sponge.

Procedure

1. In vial A (see Figure 34.1) place about 0.4 g of sodium sulfite and 0.5 mL of water. In vial B place a few drops of deionized water and 1 drop of Bromcresol Green indicator, which has a pK_a of 4.8. It will change from blue to yellow between pH 5.4 and 3.8. Add 0.5 mL of 0.01 M potassium permanganate solution to vials C, D, and E. To vials C and F

add 0.15 mL of 3 M sulfuric acid, and to vial E add 0.2 mL of 2 M sodium hydroxide solution. In vial F place a small crystal of potassium dichromate (optional). Put about 2 mL of 2 M sodium hydroxide solution in vial G. This will react with sulfur dioxide, preventing its escape into the atmosphere. Before going to the next step, take careful note of the appearance of the material in each vial.

2. Fill the reagent pipette with 1 mL of 3 M hydrochloric acid and carefully slip it into the hole in the lid of vial A. Slowly add the acid dropwise to the sodium sulfite and note carefully the changes that take place in each vial. Stop adding acid when no further change takes place in vial E.

Cleaning Up

In the hood empty the contents of all of the vials into the waste container and then rinse all of the vials, pipettes, and connecting tubes with water, which can be flushed down the drain.

Name _____ Section _____

Lab Instructor _____ Date _____

EXPERIMENT 34   Gas Chemistry: Sulfur Dioxide

PRELABORATORY QUESTION

1. Balance the equation for the reaction of sulfur dioxide with acidified dichromate ion to give chromium(III) ion.

2. Ozone, O_3, when present in the lower atmosphere, can oxidize sulfur dioxide to sulfur trioxide, which can then form sulfuric acid when it contacts water vapor. Write balanced equations for both reactions.

PHYSICAL PROPERTIES OF REAGENTS

Molecular weight of sodium sulfite   _____

Solubility of sodium sulfite   _____ g/100 mL water at 0°C

Molecular weight of sulfur dioxide   _____

Density of sulfur dioxide gas   _____ g/L

Color of potassium permanganate, $KMnO_4$ _____

Color of potassium manganate, K_2MnO_4 _____

Color of manganese chloride, iodide, or bromide, MnX_2 _____

Color of manganese dioxide, MnO_2 (insoluble in water)   _____

284   Experiment 34   Report Sheet

OBSERVATIONS

1. Appearance of material in vial A before reaction has begun   _____

   After reaction is finished   _____

   Conclusion

2. Appearance of material in vial B before reaction has begun   _____

   After reaction is finished   _____

   Conclusion

3. Chemical formulas and appearance of material in vial C before reaction has begun   _____

   After reaction is finished   _____

   Conclusion

4. Chemical formulas and appearance of material in vial D before reaction has begun   _____

   After reaction is finished   _____

   Conclusion

5. Chemical formulas and appearance of material in vial E before reaction has begun _____

   After reaction is finished _____

   Conclusion

POSTLABORATORY QUESTIONS

1. How many moles of sodium sulfite are present in this reaction if 0.40 g of the reagent is employed?

2. How many moles of hydrochloric acid are contained in 1.0 mL of a 6 M solution?

3. What volume of sulfur dioxide gas is produced at standard temperature and pressure in this reaction?

EXPERIMENT 35

Determination of Molar Mass by Vapor Density and Freezing Point Depression

Introduction

The determination of the molecular weight of a molecule is of fundamental importance in chemistry. (*Note:* The *molar mass* of a substance, the mass of one mole of molecules, was traditionally known as the *molecular weight* or, occasionally, the gram-molecular weight. Molar mass is a relatively new term, and will be used interchangeably here with molecular weight; the symbol *MW* will be used for both.) Combustion analysis may indicate that a molecule has the empirical formula CH_2; only determination of the molar mass will disclose the fact that the molecule has, for example, the molecular formula C_6H_{12} with a molecular weight of 84.16. Molar mass can be determined in more than a dozen different ways. In this experiment we will explore two of them—the vapor density method and the freezing point depression method.

A liquid that vaporizes easily under normal conditions is said to be *volatile*. The vapor density of a volatile liquid is a molar mass determination method originated by Jean Dumas in the early 1800s. It depends on Avogadro's hypothesis, which states that equal volumes of gases contain equal numbers of molecules. The procedure used in this experiment is a simplified version of more sophisticated methods used in research. At best, the method involves a fairly high degree of uncertainty (±10% error must be anticipated); even so it is adequate for those situations in which greater precision is not required. As an example, consider a case in which one is trying to determine whether a particular chlorinated hydrocarbon is monochloromethane or dichloromethane: CH_3Cl or CH_2Cl_2. Because the molar mass of dichloromethane is about 85, while that of the monochloromethane compound is only about 51, it is a simple matter to distinguish between them. On the other hand, the molar masses of cyclohexane (C_6H_{12}, MW = 84) and cyclohexene (C_6H_{10}, MW = 82) are far closer together than the ±10% uncertainty.

If this experiment were to be carried out very carefully, we would want to know the exact temperature of the gas with which we are dealing. This could be done by immersing the flask in a large beaker of water at a known temperature, for example, 100°C (boiling water). We would want to know the atmospheric pressure at the time of each run. Instead we are *assuming* in this experiment that the temperature of the gas is 400 K and that the atmospheric pressure is 1 atm. You will then calculate how far from these assumptions of 400 K and 1 atm you would need to be in order to exceed the goal of ±10% accuracy in the molar mass. Sometimes the time and effort to obtain a high degree of accuracy in an experiment are not needed.

Another method for determining molar mass depends on one of the so-called colligative properties of molecules. When a solute is dissolved in a solvent the freezing point of the solvent is lowered. The amount by which the freezing point is lowered is proportional to the number of moles of the solute dissolved in 1,000 g of the the solvent (the *molality*) and the characteristic *molal freezing point depression constant* for the solvent. A mole of a substance dissolved in 1,000 g of camphor, for instance, will depress the melting point of camphor by 37.7°C or of biphenyl by 8.0°C.

Procedure Summary

The first experiment involves filling a container of known volume with the vapor of the compound whose molecular weight is to be determined, at a known temperature and pressure. The flask is weighed empty, the liquid is added, then the flask is heated, vaporizing the sample and allowing it to more than fill the container. Air is flushed out of the flask in the process and excess vapor of the unknown escapes through a pinhole opening in the covering of the flask. After cooling, the flask is weighed again, this time with the condensed sample whose vapors just filled the flask at atmospheric pressure. The ideal gas law is used to calculate the number of moles of material present in the flask; from the mass of the sample and the number of moles present, the molar mass, in grams per mole, is readily determined.

In the second experiment the freezing point of pure biphenyl will be determined and then the freezing point will be found for a mixture containing a known mass of a compound whose molar mass is to be determined. Knowing the molal freezing point depression constant for biphenyl one can then calculate the molar mass of the unknown compound.

Biphenyl, two equivalent representations

Prelaboratory Assignment

Read the Introduction and Procedure sections and answer the Prelaboratory Questions on the Report Sheet.

Materials

Apparatus

Part 1:
Milligram balance
100-mL round-bottom flask
Flask heater (100-mL Thermowell) and controller
Aluminum foil
Copper wire
Graduated cylinder, 100 to 250 mL

Part 2:
Thermometer, preferably digital, or digital thermometer probe
Reaction tube
Thermowell sand bath

Reagents

Part 1:
Unknown volatile liquid

Experiment 35 Determination of Molar Mass by Vapor Density and Freezing Point Depression

Part 2:
Biphenyl
Unknown solid

Safety Information

1. **Safety goggles must be worn at all times in the laboratory.**
2. **The liquids used in this experiment are flammable and many are toxic by ingestion or inhalation.**
3. **Gloves are recommended for anyone with sensitive skin** (but not latex; latex is porous toward many organic liquids).

Procedure

Part 1: Molar Mass by Vapor Density Method

Carry out at least two determinations; if agreement is not within 10%, repeat until satisfactory consistency is achieved.

1. Set the controller of the flask heater to 15% power; this will heat the walls of the heater to 400 K (Figure 35.1). Weigh a 100-mL round-bottom flask, a piece of aluminum foil about 5 cm^2 and a piece of small-diameter copper wire about 15 cm long. Add about 1 or 2 mL of the unknown to the flask, then carefully fold the foil over the top of the flask, securing it to the flask with the piece of wire. Poke a small pinhole in the center of the foil.

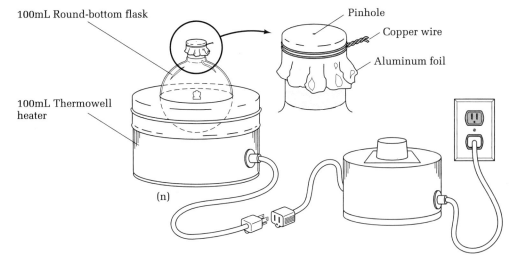

Figure 35.1 Apparatus for vapor density determination.

2. Place the flask in the well of the flask heater; the sample will boil, forcing the air from the flask, then fill the space with its own vapor. At this point, no liquid should be visible in the flask. If the liquid has a relatively high boiling point, total vaporization may be difficult to achieve and the liquid may begin to *reflux*, as indicated by rivulets of liquid running down the inside of the flask. If this happens, wrap a cone of foil (about 20 cm^2) around the top of the flask, allowing it to extend down and surround the top of the heater. This will trap the heat and better warm the neck of the flask; vaporization should be complete in about 5 to 10 minutes.

3. Remove the flask from the heater, allow it to cool to room temperature, then weigh the covered flask containing the now-condensed sample. Carefully remove the foil cap and add about 1 mL of the unknown. Re-cover the flask with the same foil piece and carry out a second trial. Calculate the mass of condensed vapor for each of your two trials. Compare the results to see if additional runs are necessary to achieve ±10% agreement.

4. When all trials have been completed, carry out the cleanup (see following Cleaning Up section). Allow the flask to air dry while you carry out Part 2 of the experiment, then continue to Part 3, the volume determination. (A 100-mL flask will contain more than 100 mL of vapor; the neck of the flask is not figured into the nominal size.)

Part 2: Molar Mass by Freezing Point Depression Method

5. Weigh a clean dry reaction tube to ±1 mg. Add to it about 1 g of biphenyl and again weigh it to ±1 mg. Melt the biphenyl on the sand bath and then insert a digital thermometer or digital thermometer probe (which is interfaced to a computer) into the molten biphenyl. Move the thermometer or probe up and down about 4 to 8 mm to stir the mixture.

    If you are using a digital thermometer record the temperature of the biphenyl at regular intervals, for example, about every 20 seconds. Continue this process until the temperature falls about five degrees after the bipheny has solidified.

    If you are using a computer with the Vernier apparatus, connect the digital temperature probe and the 9-V power supply into the serial box interface and then connect the interface box into the modem (or printer) port on the computer. Open the **Data Logger** program and in the **Experiments** folder open **Thermometer Probe** and select the **Start** button on the screen while the biphenyl is completely molten. Be sure the temperature is not above 100°C, the maximum operating temperature of the probe. Stir the mixture by moving the probe up and down while the biphenyl cools. When the temperature drops about five degrees after the biphenyl has solidified select the **Stop** button on the computer screen. You can then either write down the data from **Table A** or plot the resulting graph that is generated by the computer.

6. Carefully remove the temperature-measuring device from the solid, taking particular care not to lose any of the biphenyl that clings to the probe. Weight the tube with the remaining biphenyl to ±1 mg and then add to it about 100 mg of the unknown weighed to ±1 mg, taking care to get the unknown down on to the biphenyl and not on the upper inner walls of the tube.

7. Carefully replace the temperature probe without losing any biphenyl and heat the mixture on the sand bath. Move the temperature probe up and down in the molten material to mix the unknown thoroughly with the biphenyl. When you believe the mixture is homogeneous, allow it to cool and monitor the temperature as a function of time as before. You can repeat this melting and cooling process as often as you wish. Plot the data and determine the melting point of the mixture. Sometimes the mixture will *supercool* (cool below the freezing point without solidifying) in which case it will be necessary to determine the actual freezing point of the mixture by extrapolation (Figure 35.2).

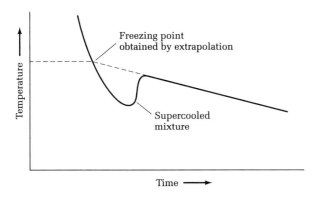

Figure 35.2 Freezing point diagram of a solution that supercools.

Part 3: Volume Determination of Flask

8. Once your final trial has been completed, waste material has been disposed of, and the flask is clean, you can measure the actual volume of the flask. To do this, fill the flask with water to the top rim of the neck, simulating the flask filled with vapor. Carefully transfer the water from the flask to a 250-mL graduated cylinder, read and record the volume to the nearest milliliter.

   Alternatively if you have access to a balance with a capacity of at least 250 g you can determine the volume of the flask by weight, bearing in mind that a gram of water has a volume of a milliliter.

Cleaning Up

Your instructor will tell you which waste receptacle to use for any liquid remaining in your flask in the first part of the experiment, and will indicate what special cleaning methods may be necessary. Once cleaning is complete and the flask is dry, carry out the volume determination (Part 3).

For the second experiment heat the tube to melt the solid and then, holding it with a test tube clamp, pour the molten solid into the container provided in the hood. Rinse out the remaining biphenyl from the tube with the minimum quantity of hexane or similar hydrocarbon solvent and pour it into the liquid waste container.

Name _____   Section _____

Lab Instructor _____   Date _____

EXPERIMENT 35   Determination of Molar Mass

PRELABORATORY QUESTIONS

1. Solve the ideal gas law for the number of moles of gas, n.

2. Write an expression for the molecular weight (MW) in terms of the sample mass, m, and the number of moles, n.

3. Write an expression for the molecular weight in terms of P, V, T, and the sample mass, m.

DATA

Part 1: Molar Mass by Vapor Density Method

	Trial 1	Trial 2	Trial 3
Mass of flask plus sample	_____	_____	_____
Mass of flask	_____	_____	_____
Mass of sample	_____	_____	_____
Mass of flask filled with water	_____	_____	_____
Mass of flask	_____	_____	_____
Mass (volume) of water	_____	_____	_____

Part 2: Molar Mass by Freezing Point Depression Method

	Trial 1	Trial 2	Trial 3
Mass of tube plus biphenyl	_____	_____	_____
Mass of empty tube	_____	_____	_____
Mass of biphenyl	_____	_____	_____
Mass of tube plus most of the biphenyl plus unknown	_____	_____	_____
Mass of tube plus most of the biphenyl	_____	_____	_____
Mass of unknown	_____	_____	_____

CALCULATIONS

All calculations are to be shown in the report.

Part 1: Molar Mass by Vapor Density Method

1. For each trial, calculate the number of moles of vapor in the flask, using your measured volume and the ideal gas law. Assume a temperature of 400 K. The molar mass (molecular weight) of the unknown is found by dividing the mass of condensed vapor by the number of moles of the substance that you calculate to be present. Report the individual values for molar mass, as well as an average value and average deviations. If your instructor gives you a list of possible unknowns, select the one that most closely matches your results. If the empirical formula for your sample is given, determine the molecular formula.

    Show sample calculations here, along with final results. Affix additional paper as needed.

Part 2: Molar Mass by Freezing Point Depression Method

2. Molality of unknown solution (show calculations)

3. Molar mass of unknown (show calculations and assume that the molal freezing point depression constant for biphenyl is 8.00 K molal^{-1})

POSTLABORATORY QUESTIONS

(Questions 3 through 6 call for in-depth analysis; single-sentence responses will likely be inadequate. Additional paper may be added as needed.)

Part 1: Molar Mass by Vapor Density Method

1. In the first experiment assuming all measurements except temperature were made correctly, how large an error in temperature would be needed to give an error of *more than 10%* in the molar mass?

2. In the first experiment assuming all measurements except for pressure were made correctly, how far off would the pressure have to be to result in an error of more than 10% in the molar mass?

3. Based on your answers to Questions 1 and 2, discuss the validity of assuming a value of 400 K for the flask heater temperature and 1.00 atm for the atmospheric pressure in the first experiment.

4. Why is the mass of air in the flask neglected during the calculations?

5. The vapor is treated as an ideal gas during the calculations. In what way(s) does this represent an unrealistic assumption? Discuss fully. (*Hint:* Under what conditions is assumption of ideality least valid?)

6. If the time of heating is insufficient to allow complete evaporation of the sample, what will be the effect on your value for molecular weight?

Part 2: Molar Mass by Freezing Point Depression Method

7. If you assume your unknown has a molecular weight of 100, what would the error in the value of the freezing point depression need to be to cause a 10% error in the result?

EXPERIMENT 36

Solubility, Complex Ions, and Qualitative Analysis

Introduction

Chemistry is the study of the chemical and physical properties of substances and their transformations. To carry out these studies, the chemist must deal with pure substances, so a large part of a chemist's work is associated with the separation of mixtures and their subsequent purification. Separations are carried out by taking advantage of the physical and chemical properties of the substances in the mixture.

Solubility, one of those properties, is often affected by pH or the presence of other ions. For example, bismuth(III) chloride forms a cloudy emulsion in distilled water, but will dissolve quickly in strongly acidic systems. Some silver compounds, insoluble in pure water, seem to dissolve very well in water to which ammonia has been added. Could it be that silver chloride prefers a high pH? How would you test that hypothesis? Adding colorless potassium thiocyanate solution, KSCN(*aq*), to pale-yellow iron(III) chloride results in a dramatic color change, but no precipitate is formed; is this a redox reaction? Is the iron(III) any less soluble (or more soluble)? In this experiment we will try to answer these and other questions related to the phenomenon of solubility. For our purposes we will define a soluble substance as one with a molar solubility $\geq 0.10\ M$.

Silver ion and chloride ion combine to give a precipitate. When ammonia is added to this precipitate it goes back into solution. *Complex ion* formation is responsible for this behavior and is the focus of the second part of this experiment. In the example just cited, the complex ion that forms is known as diamine silver, and its formula is $Ag(NH_3)_2^+$. Since ammonia is a neutral molecule, the charge on the complex is the same as that on the silver ion itself.

Many complex ions involve transition metal ions that have vacant d orbitals available for complex formation. Since concentration is critical to your interpretation of results, all of the solutions have been carefully prepared and have precisely known concentrations ($0.100\ M$, unless otherwise noted). You must take care when mixing solutions so that you do not miss something important. The procedures are very explicit in specifying dropwise addition and thorough mixing of one reagent with another; assume that there is a reason for this and proceed accordingly.

Procedure Summary

In the first part of this experiment you will determine experimentally the solubility characteristics of eight different anions and cations. You will be given $0.20\ M$ solutions of the sodium salts of eight anions (set 1) and $0.20\ M$ solutions of the nitrate salts of eight different cations (set 2).

Experiment 36 Solubility, Complex Ions, and Qualitative Analysis

In a 96-well test plate you will mix each anion with each cation in an 8 × 8 matrix and perform 64 solubility experiments. You will be looking for formation of an insoluble or sparingly-soluble product. Because all of the solutions contain their respective test ions in 0.20 M concentration the formation of a precipitate means it must have a molar solubility of <0.10 M. In Part 2 you will investigate the formation of some common complex ions.

In the third part of the experiment you will work your way through an analysis scheme, then in Part 4 your instructor will provide you with two unknowns, containing one and two of the halides, respectively, for you to identify.

Prelaboratory Assignment

Read the Introduction and Procedure sections carefully and answer the Prelaboratory Questions on the Report Sheet.

Materials

Apparatus

96-well test plate
10 × 100-mm reaction tubes (6)
10-mL graduated cylinder
100-mL graduated cylinder
250-mL beaker

Reagents

Anion set in microtip pipettes, 0.20 M (set 1):

		Names of Anions:
$NaC_2H_3O_2$	$(C_2H_3O_2^-)$	Acetate
Na_2SO_4	(SO_4^{2-})	Sulfate
NaOH	(OH^-)	Hydroxide
Na_2SO_3	(SO_3^{2-})	Sulfite
Na_3PO_4	(PO_4^{3-})	Phosphate
Na_2CO_3	(CO_3^{2-})	Carbonate
NaCl	(Cl^-)	Chloride

Cation set in mic rotip pipettes, 0.20 M (set 2):

		Names of Cations:
$Na_2C_2O_4$	(Na^+)	Sodium
KNO_3	(K^+)	Potassium
$Ca(NO_3)_2$	(Ca^{2+})	Calcium
$Fe(NO_3)_3$	(Fe^{3+})	Iron
$Cu(NO_3)_2$	(Cu^{2+})	Copper
$Al(NO_3)_3$	(Al^{3+})	Aluminum
$Pb(NO_3)_2$	(Pb^{2+})	Lead
$Zn(NO_3)_2$	(Zn^{2+})	Zinc
$Co(NO_3)_2$	(Co^{2+})	Cobalt

Other reagents:

$AgNO_3$, 0.10 M	Silver nitrate
$NH_3(aq)$, 4 M	Ammonia
$HCl(aq)$, 1 M	Hydrochloric acid
KSCN, 0.10 M	Potassium thiocyanate
$Na_2S_2O_3$, 0.20 M	Sodium thiosulfate
Starch solution	
Bleach (5.25% NaOCl)	Sodium hypochlorite
Unknown halide solutions	
t-butyl methyl ether (Optional)	
Chlorine water	

Safety Information

1. **Safety goggles must be worn in the laboratory at all times.**
2. **Silver nitrate will stain skin and clothing.** Handle it carefully; clean up all spills with sodium thiosulfate ($Na_2S_2O_3$) solution.
3. **Ammonia is caustic and has a highly unpleasant odor.** Use only in well-ventilated areas or work in the fume hood.
4. **Hydrochloric acid is corrosive to skin and clothing.**
5. **Clean up unknown halide solutions spills with copious amounts of water.**
6. **Some of the reagents contain heavy metal ions.** Consult with your instructor concerning cleanup and disposal.

Procedure

Part 1: The Solubility Tests

1. In separate wells of your 96-well test plate, test 5 drops of each anion from set 1 with 5 drops of each cation from set 2. You are looking for the formation of precipitates. These will most often appear as a cloudiness in the well, sometimes rather faint, and may be white or colored. Record your results in the Data and Observations section for Part 1 on the Report Sheet, but you should also keep notes regarding color and intensity of precipitates. When in doubt about any test, repeat it using 10 drops of each cation and anion.

2. To spot the presence of a light precipitate, it may help to perform a modified Tyndall effect test, such as is used in identification of colloidal dispersions. Hold the well plate 3 to 5 cm above a line of type (such as this paper). If there is no precipitate present at all, the type will appear clear and sharp. If the type is at all blurry, there must be a precipitate in the form of finely divided solid particles in suspension.

Part 2: Complex Ions

Aluminum and Zinc Hydroxides

3. Place 10 drops of aluminum nitrate, $Al(NO_3)_3$, in a 10 × 100-mm reaction tube. Add 1 drop of NaOH, a source of hydroxide ion, and mix thoroughly by flicking the tube (Figure 36.1).

Observe the results and record them in the Data and Observations section for Part 2. Continue the dropwise addition of NaOH until a *gelatinous* (jelly-like) precipitate forms. Try to use the number of drops of NaOH delivered to determine the formula of the precipitate.

4. Keep adding NaOH to the precipitate just formed. Determine the number of additional drops of NaOH needed to redissolve the precipitate. The redissolving indicates that a complex of Al^{3+} and OH^- has formed; try to determine the most likely formula for this complex, as well as its charge.

5 and 6. Using a clean tube, repeat the Steps 3 and 4 but substitute zinc nitrate, $Zn(NO_3)_2(aq)$, for the $Al(NO_3)_3(aq)$. As before, you should get an insoluble product from the addition of hydroxide ions, and the product should redissolve as more base is added. Try to use your observations to decide on the formulas for the hydroxide precipitate and for the soluble complex that subsequently forms.

Complexes of Silver Ion, Ag^+   In the traditional qualitative analysis scheme by which various metal cations are identified, the presence of silver ion is often confirmed by the ability of aqueous ammonia and sodium thiosulfate solutions to redissolve precipitates of silver chloride. Similar tests are used to distinguish among iodide, bromide and chloride ions, since both $NH_3(aq)$ and $Na_2S_2O_3$ will redissolve AgCl, but are not equally successful at redissolving the other two silver halides.

7. Place 8 to 10 drops of 0.100 M $AgNO_3$ in a test tube and add to it 4 drops of 1.00 M HCl. This will produce the curdy white solid, AgCl, which is photosensitive and will slowly turn gray on exposure to light. Carefully and with agitation (Figure 36.1) after each drop, add 4 M $NH_3(aq)$ to the precipitate. Continue the dropwise addition until the solid redissolves; some heat may be evolved, since you are adding a base to an acid. You may also observe a bit of white smoke in the space above the surface of the liquid in the tube. Carefully record your observations.

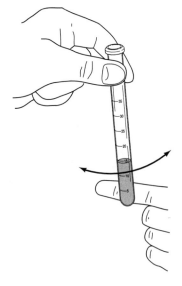

Figure 36.1 Mixing the contents of a reaction tube by flicking it. Grasp the tube firmly at the very top, and flick it vigorously at the bottom. The contents will mix without coming out of the tube.

8. Repeat the process, using sodium thiosulfate solution in place of the ammonia. Note that the concentration of $Na_2S_2O_3$ is significantly lower than that of the ammonia, so you can expect to need more of the thiosulfate to dissolve the same amount of AgCl. As before,

add the thiosulfate solution dropwise, with agitation after each drop. Note the number of drops needed to redissolve the AgCl.

Complexes of Iron(III) and Thiocyanate ions

9. In slightly acidic solutions, iron(III) ion, Fe^{3+}, also known as *ferric ion*, has a pale-yellow color. Aqueous potassium thiocyanate solution, KSCN, is colorless. Starting with 6 drops of acidified 0.10 M iron(III) nitrate in a clean tube, add 0.10 M KSCN, 1 drop at a time, mixing after each drop. Record your observations, continuing the addition until you are convinced that no further changes are occurring. (Do not exceed 15 drops of the thiocyanate.)

10. In a 250-mL beaker, dilute 1.0 mL of 0.10 M KSCN with 99.0 mL of distilled water. Using a fresh sample of 0.10 M iron(III) nitrate and a clean tube, repeat the previous experiment, using the 0.001 M KSCN you just prepared in place of the 0.10 M solution. Observe and record as before. What can you conclude? Is it possible to determine precisely the formula of the red complex ion? If not, can you make any judgments at all? For example, can you tell whether the complex contains more Fe^{3+} than SCN^-, or vice versa?

Part 3: Qualitative Analysis: The Halides

The term *halide ion* refers to any of the ions, F^-, Cl^-, Br^-, or I^- (due to its radioactivity, astatine is generally not included, notwithstanding its presence in the halogen column). You will be provided with solutions of known halide ion content and several other solutions (*test reagents*) to use in order to see how each of the halide ions reacts under a given set of conditions. By carefully keeping track of the differences and similarities in halide-ion behavior, you should be able to identify each member of the group as an unknown, and subsequently identify them even when two halide ions are present in the same solution.

Solubility with Sr^{2+} and Ag^+

11. Add 4 drops each of 0.1 M NaF, NaCl, NaBr, and NaI to four separate wells of your test plate, followed by 4 drops of 0.2 M $Sr(NO_3)_2$. Observe any changes that occur. The modified Tyndall effect test (Part 1, Step 2) may be useful here.

12. Add 2 drops of each of the halide solutions (0.1 M NaF, NaCl, NaBr, and NaI) to each of four wells in one row and immediately beneath that row again add two drops of each of the halide solutions to each of four wells. Add 2 drops of silver nitrate solution to each of the eight wells. Observe and record the results of any reactions and the appearance of any products.

13. To the top row of four wells now containing AgF, AgCl, AgBr, and AgI add 4 drops of 4 M $NH_3(aq)$ to each of the wells. Stir the mixture with a plastic toothpick or a melting point capillary; observe and record the results. Does ammonia cause the precipitate to dissolve or change color?

14. To the next row of four wells, which also contain AgF, AgCl, AgBr, and AgI, add 4 drops of 0.2 M sodium thiosulfate, $Na_2S_2O_3$. Stir and observe and record the results. What is the effect of thiosulfate on the four precipitates?

15. Finally add 4 drops each of 0.1 M NaF, NaCl, NaBr, and NaI to four separate wells of the test plate. Add 2 drops of 3% starch solution to each well. Stir and observe and record the results. Then add 2 drops of commercial bleach (5.25% NaOCl) to each well. Stir and observe and record any changes.

Part 4: Differentiation of F, Cl, Br, and I (Optional)

16. Do not use the test plate for this test because it will partially dissolve in the solvent. To further confirm the identity of the four halogens, add 4 drops of each of the 0.1 M halide solutions to small vials or reaction tubes. Then add 4 drops of *tert*-butyl methyl ether to each solution followed by 5 drops of chlorine water (water in which chlorine gas has been dissolved). Stir the mixture vigorously or shake it to mix the layers thoroughly. Observe and record any changes noted.

17. Obtain an unknown halide ion solution and test with each reagent or combination of reagents as was just done in Parts 1, 2, and 3 until the identity of the halide ion present has been determined.

18. Obtain a second unknown solution that contains two of the four halide ions, mixed together. Test with each reagent or combination of reagents until the identities of both halide ions are known.

Cleaning Up

The well plate can be cleaned by inverting it and rapping it sharply over a paper towel. The drops in each well usually come out cleanly, leaving no residue. Discard the paper towel in the container provided. If necessary, rinse residues from the wells with distilled water.

Empty the tubes from Part 4 of the experiment in the container provided. Rinse each tube with distilled water that can go down the drain.

Name _____   Section _____

Lab Instructor _____   Date _____

EXPERIMENT 36   **Solubility, Complex Ions, and Qualitative Analysis**

PRELABORATORY QUESTIONS

1. When 0.1 M NaCl is mixed with 0.1 M KCl what is the concentration of chloride ion in the resulting solution? Of sodium ion? Of potassium ion?

2. Suggest an identity for the white smoke predicted in step 7 of the procedure. Bear in mind that hydrochloric acid is an aqueous solution of hydrogen chloride gas. Write the balanced equation for formation of the product that you expect.

DATA AND OBSERVATIONS

Part 1: The Solubility Tests

					Cations				
Anions	Na^+	K^+	Ca^{2+}	Fe^{3+}	Cu^{2+}	Al^{3+}	Pb^{2+}	Zn^{2+}	Co^{2+}
1									
2									
3									
4									
5									
6									
7									

Modified Tyndall test—observations

Part 2: Complex Ions

1. Aluminum hydroxide observations

   Number of drops of NaOH needed to redissolve precipitate _____

2. Zinc hydroxide observations

   Number of drops of NaOH needed to redissolve precipitate _____

3. Silver chloride plus ammonia observations

4. Silver chloride plus thiosulfate observations

5. Complexes of iron(III) and thiocyanate ions

   Observations with 0.1 M thiocyanate

Observations with 0.001 M thiocyanate

Part 3: Qualitative Analysis: The Halides

6. Sr(NO$_3$)$_2$ reactions with

    NaF _____ NaCl _____

    NaBr _____ NaI _____

7. AgNO$_3$ reactions with

    NaF _____ NaCl _____

    NaBr _____ NaI _____

8. Addition of ammonia to the precipitates

    AgF _____ AgCl _____

    AgBr _____ AgI _____

9. Addition of thiosulfate to the four precipitates

    AgF _____ AgCl _____

    AgBr _____ AgI _____

10. Effect of mixing the four halides with starch solution and bleach

    NaF _____ NaCl _____

    NaBr _____ NaI _____

Part 4: Differentiation of F, Cl, Br, and I (Optional)

11. Test of known halides with chlorine water and *t*-butyl methyl ether.

    NaF _____ NaCl _____

    NaBr _____ NaI _____

12. Identification of an unknown halide. Outline your procedure before you perform the tests, then record your observations.

13. Identification of an unknown mixture of two halides. Outline your procedure before you perform the tests.

POSTLABORATORY QUESTIONS

1. Suggest a formula for the complex formed in Part 2, Step 7, based on the number of drops needed.

2. Write net ionic equations for the reactions that occur (if any) between the following pairs of reactants.

   a. 0.20 M strontium nitrate and 0.10 M sodium fluoride

   b. 0.10 M strontium nitrate and 0.10 M sodium chloride

   c. 0.10 M silver nitrate and 0.10 M sodium fluoride

   d. 0.10 M silver nitrate and 0.10 M sodium chloride

3. Refer to the reaction in 2(d), above. Suppose nitric acid were to be added dropwise to the tube containing the mixture after the precipitate has been redissolved by ammonia. Predict the effect of this acidification and defend your prediction.

4. What differences were you able to detect between the silver halide precipitates that would help you to confirm which was which?

EXPERIMENT 37

Water Softening

Introduction

Hard water contains calcium ions and magnesium ions, as well as ferric ions. Removing these ions is known as *water softening*. In certain parts of the United States, especially the Midwest, water percolates through layers of limestone (calcium carbonate, $CaCO_3$) to aquifers before being tapped by wells for drinking and other purposes. The resulting water is a saturated solution of calcium carbonate. The solubility of calcium carbonate in water is not high (0.0014 g per 100 mL of water), it has no taste, and the human body requires a certain amount of calcium ion to build bone. However, this water, known as "hard" water, has certain disadvantages. In particular, it forms an insoluble precipitate with soap, which results in a scum, or a "ring around the bathtub." The calcium carbonate can also form deposits on the insides of pipes, gradually closing them off.

To convert hard water to "soft" water these ions must be removed. There are several ways to do this. Nature does it by evaporation. When water evaporates the metal ions are left behind. Rain water is very soft and in days gone by was collected for such purposes as washing hair and fine clothes. Another way to remove these ions is to exchange them with other ions, in particular, hydrogen ions or, more commonly, sodium ions. This is done by passing the hard water through a bed of an ion-exchange resin.

The ion exchange resin is made of very small beads of polystyrene that have been reacted with sulfuric acid to attach sulfonic acid groups, SO_3H, to the molecule. Polystyrene is the clear brittle plastic from which disposable drinking glasses and champagne glasses are made. In foamed form it is used to make lightweight disposable insulated coffee cups. The hydrogen on the sulfonic acid group ionizes and can be replaced with other ions in an equilibrium process. If hard water is poured through the resin, the hydrogen ions are replaced by calcium ions, and the water coming from the resin bed will be acidic. To avoid this acidification, sodium chloride solution is passed through the bed and the hydrogen ions are replaced by sodium ions. Now when the hard water passes through the resin, the sodium ions are replaced by calcium ions and the water then has an excess of sodium ions.

$$2\ \text{Resin–}SO_3^-Na^+ + Ca^{2+} \rightarrow (\text{Resin–}SO_3^-)_2 Ca^{2+} + 2\ Na^+$$

The water that comes out of the resin bed is not classified as hard, although it has a high concentration of sodium ions; it is essentially soft. While the problem of acidic water is thus eliminated, the sodium ions themselves can be unhealthy for those with certain types of heart ailments.

Strong acid ion-exchange resin
Dowex 50 = ×8

Another way to soften water is by treatment with Calgon™, a water softening agent. Calgon is typical of commercially available products; the active ingredient is sodium hexametaphosphate, $Na_6P_6O_{18}$, which dissociates into sodium ions and the hexametaphosphate ions, Na^+ and $P_6O_{18}^{6-}$. This very large, very negative $P_6O_{18}^{6-}$ ion attaches itself strongly to calcium ions, trapping them in solution as a complex ion:

$$2\ Ca^{2+}(aq) + P_6O_{18}^{6-}(aq) \rightarrow Ca_2P_6O_{18}^{2-}(aq)$$

The Calgon also contains sodium carbonate, which dissolves to form sodium ions and carbonate ions; the carbonate ions remove some of the calcium ions from the water by precipitation:

$$Ca^{2+}(aq) + CO_3^{2-}(aq) \rightarrow CaCO_3(s)$$

Sodium carbonate is still sold in grocery stores under the name *washing soda* because its ability to remove the water–hardening calcium, magnesium, and ferric ions helps laundry soap clean more effectively. Although products such as Calgon may be helpful for cleaning purposes, consider whether you would like your drinking water to be softened in this manner. Water for many municipal systems is filtered through sand in order to purify it. You will run tests to determine if this also softens the water.

You will test your softened water in two ways. The first method involves a precipitation test. In the second test, you will examine two effects of the addition of soap to the treated water. First, you will note how cloudy the water becomes when the soap is added. The technical name for this cloudiness is *turbidity*. Second, you will agitate the soapy water and compare the height of the suds that forms with each sample (soap will not form a suds "head" in the presence of significant amounts of dissolved calcium ions, so the more suds you get, the softer the water must be).

Procedure Summary

Four samples of hard water are filtered through sand, a bed of ion-exchange resin, and a piece of cotton. A fourth sample is treated with a water-softening agent, Calgon. These four samples are tested with sodium carbonate and soap in order to determine the effectiveness of ion-exchange resin and Calgon for the softening of hard water.

Prelaboratory Assignment

Read the Introduction and Procedure sections carefully and answer the Prelaboratory Question on the Report Sheet.

Materials

Apparatus

Pasteur pipettes containing fine sand (1), ion-exchange resin (1), and cotton (2)
10 × 100-mm test tubes, culture tubes, or reaction tubes (8)
Corks for tubes

Reagents

Artificial hard water or local tap water if it is hard
Calgon-treated hard water
Sodium carbonate test solution, 0.1 M Na_2CO_3
Powdered soap, such as Ivory™, castile soap, or liquid Ivory™ soap, Tide™ or similar laundry detergent

---

Safety Information

1. **Safety goggles must be worn at all times in the laboratory.**
2. **Pasteur pipettes break easily, especially at the tip.** Handle them with care and notify your instructor if one is broken. Clean up broken glass carefully.

Procedure

1. Make filtration pipettes (Figure 37.1) by placing a small wad of cotton into each of four Pasteur pipettes using a wood boiling stick to push the cotton down. Fill the first about two-thirds full of sand, fill the second about two-thirds full of ion-exchange resin, and leave the last two as is.

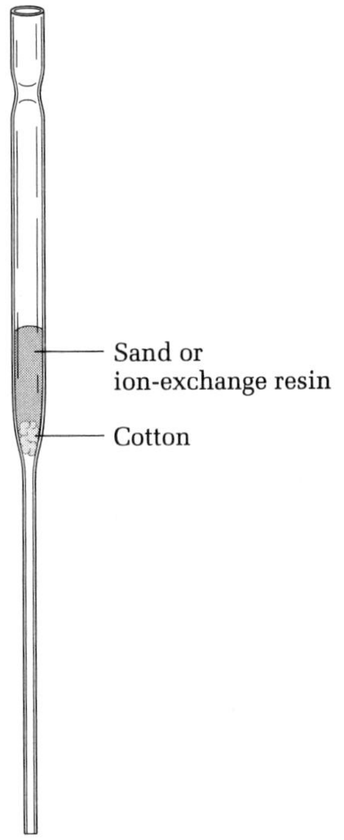

Figure 37.1 Filtration pipette.

Labels on figure: Sand or ion-exchange resin; Cotton.

2. Treat 2 mL of the artificially hard water with Calgon. Follow the directions on the box to decide the appropriate amount to add (a very small amount).

3. Label four 10 × 100-mm reaction tubes as follows: *sand*, *resin*, *Calgon*, and *cotton filter*. Place the tubes in separate wells of a 24-well test plate. Filter about 2 mL of the artificial hard water through the sand filter and collect the filtrate in the tube labeled *sand*. Transfer the water with another Pasteur pipette. The water depth will be about 2 cm. Similarly filter about 2 mL of the hard water through the Pasteur pipette containing ion-exchange resin. Filter into the third tube about 2 mL of the Calgon-treated hard water and collect the filtrate in the tube labeled *Calgon*. Through the cotton in the last Pasteur pipette filter 2 mL of the hard water into the fourth tube labeled *cotton filter*. This last tube acts as a blank. This tube will ensure that the effect of cotton on all samples is the same (we would not expect the cotton to have an effect on the results).

4. Divide the filtrates in half in the test plate, taking care to note which filtrate tubes are in which wells of the 24-well test plate. Test one portion of each sample with 3 drops of 0.1 M Na_2CO_3 solution. In the Data Table on the Report Sheet, record any evidence of precipitate formation. Compare the four filtrates so that your observations can include the relative amounts of precipitate formed.

5. To the second portion of each filtrate, add a small amount of powdered soap (about the volume of a grain of uncooked rice). Agitate the tube gently to allow the soap to mix with the water, but do not shake it; you don't want suds at this point. Compare the *turbidity* (cloudiness) of the four filtrates; the better the soap is able to mix with the water, the more turbid the sample will appear. If soap is to clean effectively, it must be able to mix with the water, so a higher degree of turbidity is a prediction of greater cleaning effectiveness.

6. Now stopper and shake the four tubes with the soap in them for about 15 seconds each. Note the relative heights of the suds that form.

Optional Extension

Repeat the four filtrations, but collect only about 1 mL of each filtrate, since you will not be doing the sodium carbonate test. In place of the powdered soap, substitute a powdered detergent (such as Tide™). Since these detergents tend to have fairly large particles, you may need to break up the pieces a bit. This is easily done by placing a small amount in a plastic bag or between sheets of smooth-finish paper, then rolling a test tube or small jar over them.

Cleaning Up

Except for the filters and packing, there is nothing here that was not, in fact, designed to go into the septic system. The pipettes containing the sand and ion-exchange resin are to be returned to the location indicated by your instructor. The pipettes containing only cotton plugs should be discarded in the container labeled "Broken Glass Only."

Name _____ Section _____

Lab Instructor _____ Date _____

EXPERIMENT 37 Water Softening

PRELABORATORY QUESTION

1. Hard water has a number of effects other than causing ordinary soap to form an insoluble precipitate. From your experience, name some of these effects. Think of the kitchen and bathroom.

DATA TABLE

	Sand	Ion-Exchange Resin	Calgon	Cotton Filter
Reaction with Na_2CO_3 solution	_____	_____	_____	_____
Turbidity with soap	_____	_____	_____	_____
Height of suds	_____	_____	_____	_____

CONCLUSIONS

1. Based on your results, which method or methods of water softening seemed most effective. Refer to your Data Table to justify your choice(s).

2. Briefly discuss some of the limitations of the method(s) that you found to be most effective in softening water. That is, for what purposes would it/they be desirable, and in what instances would the method(s) not be satisfactory?

3. Which method(s) were least effective? Suggest reasons for this ineffectiveness.

4. Summarize the connection between the amount of hardness-producing ions (such as Ca^{2+}) in the water and the cleaning ability of soap.

5. Products such as Calgon are supposed to prevent the formation of soap scum, often called "bathtub ring." Explain why such a claim might be valid.

6. Suggest a reason why the soap tests could not be done using the same filtrate samples that were treated with sodium carbonate solution. (*Hint:* Refer to the Introduction.)

EXPERIMENT

38 Growing Crystals In Gels

Introduction

Crystallization plays a major role in chemistry. In organic chemistry it is an important method of purification. Two common substances, salt and sugar, are sold as very pure crystals. In former times most salt was obtained from seawater by slow evaporation, a process still carried out at the south end of San Francisco Bay. Sugar crystals are obtained by cooling a hot, saturated solution. Many inorganic crystals can be obtained by slow evaporation of the solvent, usually water.

When two soluble salts are mixed to give an insoluble substance (for example, when colorless lead acetate solution is mixed with colorless potassium iodide solution), there is an immediate precipitate of yellow lead iodide along with the formation of soluble potassium acetate:

$$Pb(OCOCH_3)_2(aq) + 2\ KI(aq) \rightarrow PbI_2(aq) + 2\ KOCOCH_3(aq)$$

This precipitation reaction gives lead iodide crystals that are microscopic in size, but if they could be grown slowly they would be large and in many cases quite beautiful. This can be done by allowing the two reactants to come in contact with each other very slowly, a process most easily carried out in a gel. Even crystals of metals can be formed in this way.

Under certain circumstances instead of crystal growth a precipitate forms. Diffusion can then take place through this precipitate until a critical concentration is again reached, at which time another precipitate will form beyond the first. The result is a series of rings of precipitate, called Liesegang rings, after their discoverer. It is this phenomenon that is thought to account for the bands in certain minerals, such as banded agate.

Silica gel is a three-dimensional network, a polymer, made up mostly of water. One reactant dissolved in this water remains fairly stationary. The second reactant, added as a solution to the top of the gel, diffuses (a slow process) into the gel to react slowly with the first to form crystals.

The technique of growing crystals in gels is simple and fascinating. Special supplies and equipment are not necessary, and the method described affords an opportunity to grow near-perfect crystals. Water glass (sodium metasilicate, Na_2SiO_3) solution will form a gel when acid is added to it carefully. The acid (acetic acid, CH_3COOH) dissolves one of the crystal-forming reagents; a solution of the second component of the reaction is added to the surface of the gelled water glass and eventually crystals form in the gel. Other gels such as agar or gelatin provide different opportunities for carrying out this experiment, but silica gels seem to favor good crystal formation, whereas organic gels favor amorphous or microcrystalline deposits.

Silica gels are made by precipitating unstable silicic acid, H_4SiO_4, from the sodium silicate solution. The silicic acid molecules react with each other forming siloxane bonds, O–Si–O, and splitting out water to form a three-dimensional network. (See Experiment 42,

Synthesis of Slime, for a discussion of the structure of silicic acid.) The acetic acid is added in excess, resulting in a system that is buffered by the acetic acid/acetate ion system, CH_3COOH/CH_3COO^-.

Most acetates are soluble, and a low pH inhibits the precipitation of basic salts or silicates. The standard gel is distinctly on the acid side of the buffer region of the acetic acid/acetate system. The more slowly the gels set, the more transparent they are likely to be. Incorporating one reagent into the acetic acid before mixing with the water glass avoids precipitation of silicates or hydroxides.

Procedure Summary

In each of the following experiments a dilute solution of sodium metasilicate (water glass) is mixed with acetic acid into which an ionic substance has been dissolved. The gel is allowed to form overnight, then a metal ion or other cation is placed on the surface of the hardened gel. Crystals of the new, insoluble substance form in the gel as the top aqueous ion diffuses downward into the gel.

Prelaboratory Assignment

Read the Introduction and Procedure sections carefully and answer the Prelaboratory Questions on the Report Sheet.

Materials

Apparatus

Pasteur pipette
Pipette, 1.0×0.01 mL
Culture tubes, 10×100 mm

Reagents

Sodium silicate solutions:
    Standard solution *(See Prelaboratory Question #1)*
    For Gel #10: 9.0 mL of saturated sodium silicate + 21.0 mL of distilled water
Acetic acid: 1 M
        0.5 M (prepared by diluting 1 M $HC_2H_3O_2$ with an equal volume of distilled water)
        0.6 M (prepared by diluting 3 parts of 1 M $HC_2H_3O_2$ with 2 parts of distilled water)

Solutions of the reactant ions:
    1.0 M $CuCl_2$
    1.0 M NaCl
    1.0 M lead acetate, $Pb(C_2H_3O_2)_2$
    2.0 M KI
    1.0 M KI (prepared by diluting 2.0 M KI with an equal volume of distilled water)
    3.0 M Tartaric acid, $HOOC(CHOH)_2COOH$
    Saturated solutions of
        KNO_3
        $CuSO_4$
    1.0 M $CuSO_4$
    1.0 M hydroxylamine hydrochloride, $NH_2OH \cdot HCl$
    0.10 M Hydrazine hydochloride, $N_2H_4 \cdot HCl$
    1.0 M Potassium chromate, K_2CrO_4
    1.0 M $AgNO_3$
    0.5 M $AgNO_3$ (prepared by diluting 1.0 M $AgNO_3$ with an equal volume of distilled water)

Optional Reagents (needed only if gels #12 and 13 A-D are to be made)

    Saturated mercuric chloride solution, Hg_2Cl_2
    1.0 M mercuric nitrate, $Hg(NO_3)_2$
    Saturated sucrose solution, $C_{12}H_{22}O_{11}$
    1.0 M $HC_2H_3O_2$, saturated with sodium chloride, NaCl
(All other reagents for these gels are used in one or more of the preceding combinations.)

> **Safety Information**
> 1. **Safety goggles must be worn at all times in the laboratory.**
> 2. **The heavy metal ions are toxic.** Handle them with care and wash your hands before leaving the laboratory (always a good practice).

Procedure

General Procedure

1. Measure 1.5 mL of 1.0 M acetic acid and place it in a 10×100-mm culture tube.
2. Measure the aqueous ion listed for the particular experiment and add it to the acid, mix well by stoppering the tube and inverting it several times.
3. Add 1.5 mL of dilute sodium metasilicate solution to the acid/cation mixture. Stopper the tube and mix the contents well. Setting time will vary from a few minutes to several hours or even days, depending on conditions and specific reagents used.
4. When the gel mixture has set, carefully add the developing metal or solution to the top of the gel, stopper the tube, and place it in a safe display area. Crystals will develop slowly over many days.

Experiment 1: Copper Tree

The gel: 0.3 mL of 1.0 M $CuCl_2$
Iron brad, wire or very small nail, ungalvanized
0.3 mL 1.0 M NaCl
Development: Add the copper solution to the acetic acid. After the gel sets, gently push an ungalvanized iron brad or small nail or iron wire into it. Add the sodium chloride solution on top of the nail.

Experiment 2: Lead Tree

The gel: 0.3 mL 1.0 M $Pb(CH_3COO)_2$ or 1.0 M $Pb(NO_3)_3$
Zinc metal, 0.3×1.0 cm
Development: Add the lead solution to the acetic acid. After the gel sets, gently insert a thin piece of zinc metal, about 0.3×1.0 cm.
To prevent dehydration, add 6 to 7 drops of distilled water followed by 3 to 4 drops of 1 M acetic acid.

Experiment 3: Lead Iodide Ferns

The gel: 0.3 mL 1.0 M Pb(CH$_3$COO)$_2$
0.6 mL 2.0 M KI
Development: Add the lead solution to the acetic acid. After the gel sets, add the potassium iodide solution.

Experiment 4: Lead Iodide

The gel: 0.12 mL 1.0 M KI
0.6 mL 1.0 M Pb(CH$_3$COO)$_2$
Development: This is just the reverse of the previous procedure; add the KI solution to the acetic acid. After the gel sets, develop with the lead acetate.

Experiment 5: Potassium Tartrate

The gel: 1.5 mL 3.0 M tartaric acid, HOOC(CHOH)$_2$COOH (used in place of acetic acid)
Saturated KNO$_3$ or KCl
Development: After the gel sets (4 to 5 days in this case), add 0.6 mL of the KNO$_3$ carefully by pipette.

Experiment 6: Copper(II) Tartrate

The gel: 1.5 mL 3.0 M tartaric acid, HOOC(CHOH)$_2$COOH (used in place of acetic acid)
Saturated CuSO$_4$
Development: After the gel sets (4 to 5 days), add 0.6 mL of the copper sulfate solution carefully and slowly.

Experiment 7: Copper Crystals

The gel: 0.12 mL 1.0 M CuSO$_4$
1.0 M hydroxylamine hydrochloride, NH$_2$OH•HCl
Development: Add the copper sulfate to the acetic acid solution. When the gel has set, add 0.6 mL of the hydroxylamine.

Experiment 8: Copper(II) Reduction with Hydrazine Hydrochloride

The gel: 0.12 mL of 1.0 M CuSO$_4$
0.1 M N$_2$H$_4$•HCl
Development: Add 2.0 mL of the copper sulfate to the acetic acid. When the gel has set, add 10 mL of the hydrazine. Nitrogen gas, N$_2$, is formed, creating flat bubbles in the silica gel.

Experiment 9: Copper(II) Chromate/Copper(II) Hydroxide

(Note that this system uses a different acetic acid concentration.)
The gel: 0.12 mL of 1.0 M K_2CrO_4
1.0 M $CuSO_4 \cdot 5\ H_2O$
Development: Add 2.0 mL of the chromate solution to 25 mL of 0.5 M acetic acid. When the gel has set, carefully add 0.9 mL of copper sulfate solution; a true Liesegang ring system results.

Experiment 10: Silver Chromate

(Note that this system uses different silicate and acetic acid concentrations.)
The gel: 0.25 mL of 1.0 M K_2CrO_4
1.5 mL of 0.6 M acetic acid (note the concentration)
0.45 mL of sodium silicate (9.0 mL saturated solution + 21.0 mL water; note the concentration)
0.3 mL 1.0 M $AgNO_3$
Development: Add the potassium chromate to 1.5 mL of 0.6 M acetic acid. Now add 0.45 mL of the special sodium silicate solution. When the gel has set, add 0.3 mL of the silver nitrate solution. Try the experiment again, this time using standard silicate solution and 1.0 M acetic acid; the results are quite different.

Experiment 11: Silver Tree

The gel: No additives
Development: Carefully place a piece of aluminum foil in a vertical position in the bottom of the culture tube. Add the gel components. After the gel has set, add 1.0 mL of 0.5 M silver nitrate solution to the top of the gel.

NOTE: If your institution has the means to safely dispose of toxic mercury waste the following two experiments are interesting to perform.

Experiment 12: Mercury(II) Oxide/Mercury(II) Chloride

The gel: Saturated $HgCl_2$ (Caution! Mercury compounds are toxic.)
1.5 mL of 0.5 M acetic acid added to 1.5 mL of the standard silicate solution.
No other ions are incorporated. This is a rapid-setting gel.
Development: After the gel is set, add 0.9 mL of the saturated $HgCl_2$ solution.

Experiment 13: Mercury Iodide (Four Systems)

A. *The gel:* 1.0 M KI
1.0 M $Hg(NO_3)_2 \cdot 2\ H_2O$ (Caution! Mercury compounds are toxic.)
Development: Add 0.15 mL of the potassium iodide solution to the acetic acid. Once the gel has set, add 0.9 mL of the mercuric nitrate solution.
Do this for two tubes.

This gel (A) illustrates the effect of light on a chemical reaction (photochemistry). Prepare two of the tubes as above, wrap one in foil or heavy paper; unwrap only briefly, to observe progress. The unwrapped tube will undergo photochemical decomposition. What substance is formed?

B. *The gel:* 1.0 M KI
1.0 M Hg(NO$_3$)$_3$•2 H$_2$O (Caution! Mercury compounds are toxic.)
Saturated sucrose solution
Development: As for A, but add 0.15 mL of the saturated sucrose solution to the mixture of sodium silicate and acetic acid.

C. *The gel:* 1.0 M acetic acid saturated with NaCl
1.0 M KI
Development: Mix 1.5 mL of the saline acetic acid with 0.15 mL of potassium iodide solution, then add the 1.5 mL of standard sodium silicate. In this system, the mercury is converted to a chloro complex.

D. *The gel:* Solid KI
0.4 M HgCl$_2$
Development: Add 1.0 mL of the sodium silicate solution to 1.0 mL of 1 M acetic acid in which between 15 and 150 mg of potassium iodide has been dissolved (record the amount). After the gel sets, add 1.0 mL of 0.4 M mercuric chloride solution. The appearance of the final gel will depend on the potassium iodide concentration.

Cleaning Up

The crystals formed in these experiments are insoluble in water. The gel is not toxic and the culture tubes are relatively inexpensive, so it is suggested that, except for the mercury-containing gels, the tubes be disposed of in the nonhazardous solid waste. The mercury-containing gels must be disposed of in the hazardous solid waste container. Alternatively, the gels can be scraped from the culture tubes and placed in the appropriate containers and the tubes washed and reused.

EXPERIMENT 38   Growing Crystals in Gels

PRELABORATORY QUESTIONS

1. Calculate the sodium metasilicate concentration in the solution you are using to prepare gels (before adding acetic acid). Na_2SiO_3 is sold as a solution that the catalog says has a SiO_2 content of 45% (presumably weight/volume). The reagent you will use has been prepared by diluting 150 mL of this saturated solution to 1 L. What is the weight percent of SiO_2 in the final gel after the acetic acid has been added?

2. In the *Handbook of Chemistry and Physics,* look up the solubilities and colors of the reagents and products for the crystals that you will prepare. Be prepared to use this information as you write down your observations and conclusions.

Experiment 38   Report Sheet

CALCULATIONS AND OBSERVATIONS

Experiment 1: Copper Tree

Write the balanced equation for the reaction

Observations and conclusions

Experiment 2: Lead Tree

Write the balanced equation for the reaction

Observations and conclusions

Experiment 3: Lead Iodide Ferns

Write the balanced equation for the reaction

Observations and conclusions

Experiment 4: Lead Iodide

Write the balanced equation for the reaction

Observations and conclusions

Experiment 5: Potassium Tartrate

Write the balanced equation for the reaction

Observations and conclusions

Experiment 6: Copper(II) Tartrate

Write the balanced equation for the reaction

Observations and conclusions

Experiment 7: Copper Crystals

Write the balanced equation for the reaction

Observations and conclusions

Experiment 8: Copper(II) Reduction with Hydrazine Hydrochloride

Write the balanced equation for the reaction

Observations and conclusions

Experiment 9: Copper(II) Chromate/Copper(II) Hydroxide

Write the balanced equation for the reaction

Observations and conclusions

Experiment 10: Silver Chromate

Write the balanced equation for the reaction

Observations and conclusions

Experiment 11: Silver Tree

Write the balanced equation for the reaction

Observations and conclusions

Experiment 12: Mercury(II) Oxide/Mercury(II) Chloride

Write the balanced equation for the reaction

Observations and conclusions

Experiment 13: Mercury Iodide (4 systems)

Write the balanced equation for the reactions

    A. _____

    B. _____

    C. _____

    D. _____

Observations and conclusions

POSTLABORATORY QUESTIONS

1. Speculate on the effect of temperature on the reactions described in this experiment. Would an increase in temperature increase the rate of formation of the various crystals and precipitates? What would the effect of an increase in temperature have on crystal size?

2. Speculate on the appearance of some of these experiments if they were conducted in a weightless environment such as exists in outer space.

3. Cite examples of the diffusion of:

   a. a gas within a gas.

   b. a solid within a gas.

   c. a liquid within a liquid.

EXPERIMENT 39

Synthesis and Analysis of Aspirin

Introduction

To a chemist, molecules—especially the larger ones—are most easily thought of in terms of the functional groups they contain. These functional groups are clusters of atoms that are found to occur in the same sequence in many different compounds. Just as simple molecules are viewed as being made up of individual atoms, more complex molecules are often identified by the functional groups that they contain. Some typical groups are shown below; notice that some are neutral (and have a *-yl* ending to their names) while others are charged

—OH
Hydroxyl group

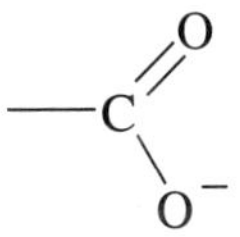

Carboxylate anion

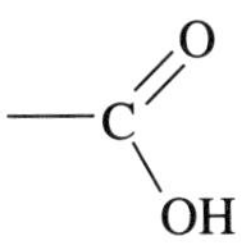

Carboxyl group

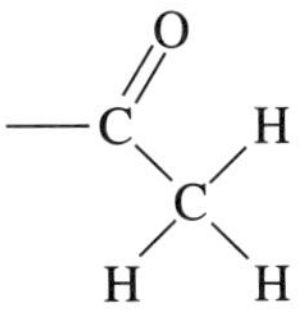

Acetyl group   which is the same as

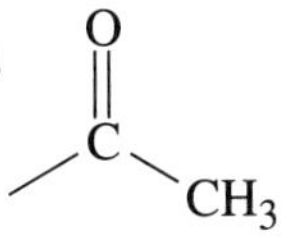

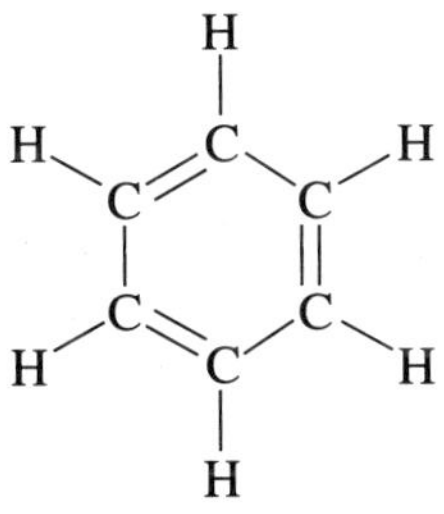

Benzene

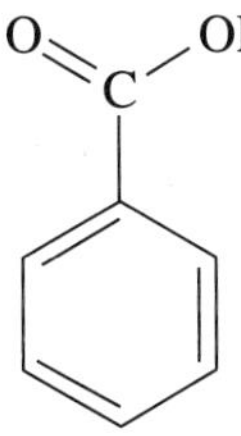

Benzoic acid   which is the same as

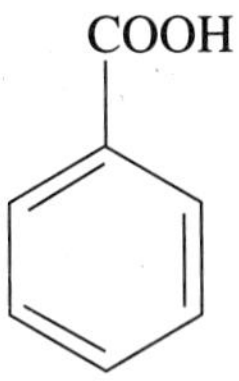

You can see from the formulas above that certain shortcuts (hexagon with alternating double/single bonds) quickly appear in structure writing, three of which are seen in the structure shown for benzoic acid. First, the corners of the hexagon are understood to be occupied by carbon atoms. Second, rather than draw out the entire carboxyl group, it can be indicated by

—COOH, which assumes the reader knows that the first oxygen atom is doubly bonded to the carbon atom (called a carbonyl carbon), whereas the second oxygen is the one between the carbon and the hydrogen. In addition, the hydrogens have been left off the ring; the substituent is shown only for those sites at which the carbon is bonded to some atom or group other than hydrogen.

Salicylic acid + Acetic anhydride ⟶ Acetylsalicylic acid (aspirin) + Acetic acid

Procedure Summary

In the experiment that follows, you will synthesize aspirin from salicylic acid and acetic anhydride. The other product from this balanced equation is acetic acid. You will then test the purity of your aspirin by a technique known as colorimetric analysis.

Prelaboratory Assignment

Read the Introduction and Procedure sections and answer the Prelaboratory Questions on the Report Sheet.

Materials

Apparatus

Reaction tube with cork
   or
10 × 100-mm culture tube with cork
Boiling stick
Steam bath or boiling water bath
96-well test plate (flat-bottom preferably)

Reagents

Salicylic acid
Acetic anhydride
Concentrated phosphoric acid
Distilled water
50% aqueous ethanol (solvent for Part 2)
Iron(III) chloride test solution

Safety Information

1. **Safety goggles must be worn at all times in the laboratory.** Also wear an apron.
2. **Work in the hood while handling acetic anhydride and phosphoric acid.** Acetic anhydride has an extremely harsh and unpleasant odor. Avoid inhaling vapors from the acid mixture.
3. **Salicylic acid dust can be an irritant.**

Part 1: Synthesis of Aspirin

1. Weigh out (±1 mg) a sample of salicylic acid between 50 and 70 mg in a reaction tube or 10 × 100-mm culture tube.
2. This step should be done under the hood. Taking care to keep it off your skin, add 15 drops of acetic anhydride and 3 drops of concentrated (85%) phosphoric acid to the salicylic acid in your test tube. Return to your work area, and continue the procedure.
3. Heat the mixture on a steam bath or in a beaker of very hot (90°C) water for 5 minutes to complete the reaction.
4. Cautiously add 10 drops of water, cork the tube, and shake well. Add 10 more drops of water and then cool the tube in an ice bath. Crystals should begin to form as the mixture cools; if they do not, try scratching the inside of the tube with a glass stirring rod. Use a Pasteur pipette to remove as much liquid as possible from your crystalline product. To do this, blow the air from a pipette as you push it through the crystals to the bottom of the tube. Gently release the rubber bulb to suck the liquid into the pipette. Filtration occurs between the tip of the pipette and the bottom of the tube (Figure 39.1).
5. To recrystallize the aspirin, add just enough water to cover the crystals, then heat the tube on a steam bath or a hot water bath as before. Heating should bring the sample into solution. Do not waste time during this procedure; prolonged heating of aspirin with water will decompose it. Add more water only if it is needed; by no means should the total exceed about 0.5 mL. Allow the tube to cool spontaneously to room temperature and then cool it in an ice bath. Once more, remove as much water as possible from the crystals then scrape the crystals onto a piece of filter paper and squeeze them between sheets of filter paper to remove the water. Once the product is dry determine its mass and, if so instructed, its melting point. It would not be a good idea to ingest your aspirin should you have a headache. Never taste chemicals in the laboratory. Examine the crystalline product with a hand lens, comparing the crystal shape to that of the starting material.

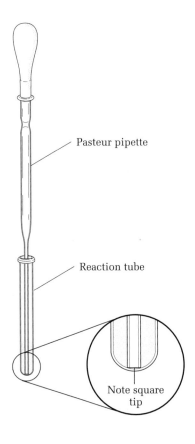

Figure 39.1 Filtration using the Pasteur pipette and reaction tube.

Part 2: Analysis of the Product

This part of the experiment involves a test of the purity of your aspirin. Because all of the other reactants and products used or formed in Part 1 are either liquids or are very soluble in water, essentially the only constituents of your dried product are aspirin and any remaining unreacted salicylic acid.

Salicylic acid will react with iron(III) ion, Fe^{3+}, in a way that aspirin will not. In alkaline (basic) solution, the free acid forms a complex ion with iron(III) that is easily identified by its violet color. Because the aqueous metal ion by itself has a faint yellow color, and because the two organic compounds give colorless solutions, detection of the complex is a simple matter of recognizing the presence of the violet color.

A quantitative determination of the amount of unreacted salicylic acid can be made by measuring the intensity of the violet color. This can be done either visually or by use of a spectrophotometer. In either case, a set of standards is prepared by making a series of dilutions of a salicylic acid solution of known concentration, then each dilution is reacted with iron(III) ion. The intensity of color varies with the concentration of salicylic acid. A solution of the product from Part 1 is then tested with iron(III), and the intensity of its color is compared with the range of intensities of the standards.

Both aspirin and salicylic acid have limited water solubility. For that reason, a cosolvent system is needed for the test. A solution containing approximately 100 mg of salicylic acid in 100 mL of a 50 : 50 mixture of ethanol and water has been prepared for you; the precise concentration is given on the container label, and should be recorded for use in the calculations to be done later. There is also a prepared solvent mixture for use in dissolving your solid product for analysis.

6. Prepare a solution of your product from Part 1 by dissolving about 20 mg (±1 mg) of your dry product in 20.0 mL of 50 : 50 water/ethanol solvent mixture. Allow the solid to dissolve completely before proceeding.

7. Place 1 drop of the standard salicylic acid solution in well B-1 of your 96-well test plate, 2 drops in B-2, 3 in B-3, and so on through B-10. Now, starting with B-9, and working down, add 1 drop of distilled water to B-9, 2 drops of water to B-8, 3 drops to B-7, etc., until all 10 wells contain 10 drops of liquid each (well B-10 contains only salicylic acid solution). Add 1 drop of standard iron(III) test solution to each of the wells in row B and stir each with a toothpick, preferably plastic, or a glass melting point capillary.

8. Place 10 drops of the solution you prepared from your product in each of the first 10 wells of row A. Add 1 drop of standard iron(III) test solution to each well of row A, and determine the relative concentration of salicylic acid in your sample by comparing the color caused by addition of Fe^{3+} to your solution with those colors obtained with the standards in row A.

    If your material seems to be intermediate between two of the standards, report it as such. If your product gives a more intense color than the mixture in well A-10, you have converted less of the salicylic acid to aspirin than expected. Make a 1 : 10 dilution of your product solution and repeat the comparison, with 10 drops of the diluted mixture in each of the first 10 wells of row C. As before, add 1 drop of the ferric ion test solution to each of the wells C-1 through C-10, and compare with row B, as before. Comparisons are most easily made by holding the test plate about 5 to 10 cm above a sheet of white paper in a well-lighted area, then looking straight down through the wells.

9. Find an old bottle of aspirin, the older the better. Test it for salicylic acid in the same way (see the Postlaboratory Question).

The ferric chloride test is a general test for molecules that contain the phenol group.

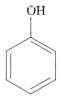

Phenol

Cleaning Up

Everything used in this experiment can be safely rinsed down the sink with large amounts of water.

Name _____ Section _____

Lab Instructor _____ Date _____

EXPERIMENT 39   Synthesis and Analysis of Aspirin

PRELABORATORY QUESTIONS

1. Examine the structures of acetic anhydride and acetic acid given in the Introduction. Look up the meaning of the term *anhydride* in your text or other reference. Explain how the definition of anhydride is appropriate in the case of acetic anhydride. (*Hint*: Think of two molecules of acetic acid.)

2. Give directions for preparing 100.0 mL of 0.100 M iron(III) chloride solution. Express the concentration of the resulting solution in milligrams of Fe^{3+} per milliliter of solution, mg Fe^{3+}/mL.

CALCULATIONS AND CONCLUSIONS

1. Calculate the theoretical yield of aspirin, based on the mass of salicylic acid you used in Part 1 of the procedure. Use this and the mass of your product to calculate your percentage yield of crude product.

2. Calculate the concentration of salicylic acid present in each of the 10 wells of row B, using the concentration given for the standard salicylic acid solution. Show your work for well B1; report values for the other nine wells.

3. Based on the values you calculated in the previous question, and the observations you made in Part 2, estimate the salicylic acid concentration in the solution made from your product. By comparison with the concentration of the solution you made from your crude product, determine the concentration of aspirin (mg/mL) and the percent purity of your aspirin product from Part 1.

4. Calculate your actual percentage yield in the aspirin synthesis by combining the figures for the yield of crude product with the percentage purity.

POSTLABORATORY QUESTION

1. Bottles of aspirin are often dated and have a "use by" date. When you open an old bottle of aspirin, it often smells like vinegar. Why?

2. When aspirin is boiled for some time with water two new compounds are formed. One has the molecular formula $C_2H_4O_2$ and the other $C_7H_6O_3$. What are the structures of these two compounds?

EXPERIMENT 40

Synthesis of Esters

Introduction

Many of the common "flavors," especially those of fruits, are actually odors. If you pinch your nose and are blindfolded you will find it difficult to distinguish an apple from a potato by taste. Many of the odors of fruits are esters that can be synthesized by combining an organic acid with an alcohol. The carboxylic acid group combines with the hydroxyl group of the alcohol to form the ester and water in the presence of an acid catalyst.

$$R-\underset{OH}{\overset{O}{\underset{\|}{C}}} + R'-OH \xrightarrow{H^+} R-\underset{O-R'}{\overset{O}{\underset{\|}{C}}} + H_2O$$

Carboxylic acid   Alcohol   Ester

In this experiment you will synthesize some esters, note their odors, and note their solubility in water.

Procedure Summary

Equimolar amounts of the organic acid and the alcohol are allowed to react in the presence of the acid catalyst. We will use both the traditional catalyst, concentrated sulfuric acid, and a safer more convenient catalyst, Dowex 50× ion-exchange resin. This catalyst is simply sulfuric acid in which one of the protons has been replaced by polystyrene. It comes in the form of tiny plastic beads and so, unlike sulfuric acid, is easily removed from the reaction mixture by simple filtration.

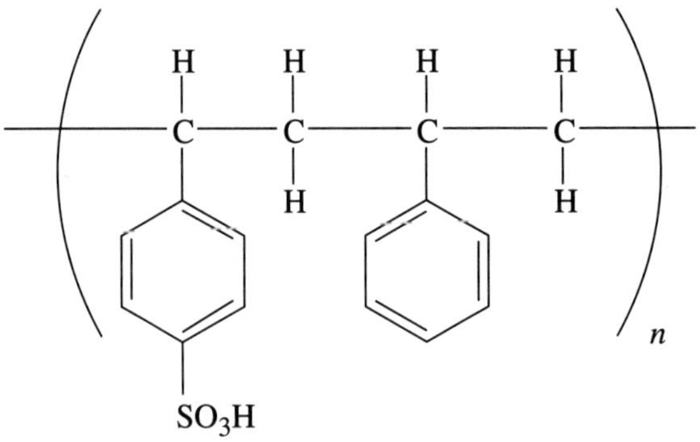

Strong acid ion-exchange resin Dowex 50X

Prelaboratory Assignment

Read the Introduction and Procedure sections carefully and answer the Prelaboratory Questions on the Report Sheet.

Materials

Apparatus

10 × 100-mm reaction tubes
   or
10 × 100-mm culture tubes
Boiling chips or wood boiling sticks
Sand bath and controller
Pasteur pipettes and rubber bulbs
1-mL graduated pipette and syringe adapter

Reagents

Methanol
1-Propanol
1-Butanol
iso-Amyl alcohol
Acetic acid
Propanoic acid
Salicylic acid
Dowex 50X ion-exchange resin
Sulfuric acid, concentrated

Experiment 40 Synthesis of Esters

Safety Information	1. Safety goggles must be worn at all times in the laboratory. 2. **Be particularly careful with concentrated sulfuric acid.** Wipe up all spills immediately. 3. **To smell these reactants and products, immerse a wood stick in the liquid and waft the odor carefully toward your nose with your hand.**

Procedure

In a 10 × 100-mm reaction tube or a 10 × 100-mm culture tube place equimolar amounts of the alcohol and the acid and 0.05 g (50 mg) of the strong acid ion-exchange catalyst, Dowex 50X. The two organic reagents can be measured in one of three ways. The most precise—and also the most time consuming— is to weigh them. The next best method, and the one recommended for this experiment, is to measure them volumetrically. The third method is to simply estimate the volumes by counting drops from a Pasteur or Beral pipette.

Compound	MW	Density	Boiling Point, °C	Formula	Melting Point, °C
Methanol	32.04	0.791	65	CH_3OH	−94
1-Propanol	60.10	0.804	97	$CH_3(CH_2)_2OH$	−126
1-Butanol	74.12	0.810	118	$CH_3(CH_2)_3OH$	−90
iso-Amyl alcohol (3-Methyl-1-butanol)	88.15	0.809	130	$(CH_3)_2CHCH_2CH_2OH$	−117
Acetic acid	61.06	1.049	116	CH_3COOH	17
Salicylic acid	138.12	1.44	—	(structure: 2-hydroxybenzoic acid)	158–160°C

Some Possible Esters

Name	MW	Odor	Formula
n-Propyl acetate	101.7	Pear	$H_3C-\overset{O}{\underset{\|}{C}}-O-(CH_2)_2CH_3$
n-Butyl acetate	126.5	Apple	$H_3C-\overset{O}{\underset{\|}{C}}-O-(CH_2)_3CH_3$

iso-Amyl acetate	142	Banana	H₃C—C(=O)—O—(CH₂)₂CH(CH₃)₂
Methyl salicylate	222	Wintergreen	2-hydroxybenzoate methyl ester (OH on benzene ring, C(=O)—O—CH₃)

The four alcohols can be reacted with an equimolar amount of the acetic acid using the Dowex resin catalyst.

1. In a typical reaction you might react 0.247 g (0.31 mL) of 1-butanol with 0.204 g (0.19 mL) of acetic acid in the presence of 50 mg of Dowex 50X to prepare *n*-butyl acetate.
2. Note the odor of the mixture.
3. Remove 2 drops of the mixture and add it to about 0.5 mL of water. Do the reactants dissolve in the water?
4. Add a boiling stone and heat the mixture to boiling on a sand bath for about 3 minutes. Be very careful not to boil the reactants out of the tube. One way to guarantee this is to hold the tube between the thumb and forefinger when carrying out the boiling. If your fingers get warm, remove the tube from the heat temporarily. The tube can also be held with a test tube holder. Be careful not to point the tube at yourself or at anyone else, because the mixture may bump (that is, boil violently) and spurt out of the tube. Note that the vapors of the liquid ascend the tube a short distance, condense, and return to the bottom of the tube. This process is called *refluxing*.
5. At the end of the refluxing period put a wood stick into the reaction mixture, withdraw it, and carefully note the odor of the mixture.
6. Using a Pasteur pipette withdraw some of the liquid (but not the resin) and add it to about 1 mL of water in a reaction tube or culture tube. Mix the contents thoroughly. Does the organic liquid dissolve in the water?

Preparation of Methyl Salicylate

Since salicylic acid is less reactive than acetic acid, a stonger catalyst, sulfuric acid, is used for this reaction.

7. Add 1 drop, no more, of concentrated sulfuric acid to a mixture of 0.4 mL of methanol (an excess) and 0.10 g (100 mg) of salicylic acid.
8. Add a boiling stone, reflux the mixture for 5 minutes, and then note the odor of the product using a wood stick as before.

Cleaning Up

Place the organic products in the waste container labeled *Flammable Waste*. Wash the tubes carefully with soap and water and allow them to dry. If necessary to speed the drying process the tubes can be rinsed with a few drops of acetone, which is added to the waste container.

Name _____ Section _____

Lab Instructor _____ Date _____

EXPERIMENT 40   Synthesis of Esters

PRELABORATORY QUESTIONS

1. In the Procedure example, 0.247 g of 1-butanol is used. How many moles is this mass of butanol?

2. In the Procedure example, 0.206 g of acetic acid is used. How many moles is this mass of acetic acid?

3. What is the volume of 3.33×10^{-3} moles of 1-propanol?

Experiment 40   Report Sheet

DATA AND OBSERVATIONS

	Ester 1	Ester 2
Alcohol used	_____	_____
Acid used	_____	_____
Name of ester	_____	_____
Mass of alcohol	_____	_____
Volume of alcohol	_____	_____
Mass of acid	_____	_____
Volume of acid	_____	_____
Original mixture odor	_____	_____
Original mixture: Solubility in water	_____	_____
After refluxing odor	_____	_____
After refluxing: Solubility in water	_____	_____

POSTLABORATORY QUESTIONS

1. Examine the chemical structure of aspirin (Experiment 39). Is it an ester? Why or why not?

2. According the the equation on the first page of this experiment ester formation is an equilibrium reaction. Often the equilibrium constant for the reaction is near one. If equimolar amounts of an acid and an alcohol were allowed to react with one another what would be the concentration of the ester in the final equilibrium mixture if the equilibrium constant were 1.0?

EXPERIMENT 41

Oxidation–Reduction: Dyeing with Indigo, the Blue Jeans Dye

Introduction

Oxidation and reduction are a pair of the most common chemical reactions. The rusting of iron (oxidation of iron by oxygen) and the manufacture of iron (reduction of iron ore by coke) are two important examples. For metals like iron, as for all elements, oxidation involves a loss of electrons and a gain in oxidation number:

$$Fe° \rightarrow Fe^{3+} + 3e^-$$

Reduction is a gain of electrons and a decrease in oxidation number:

$$Fe^{3+} + 3e^- \rightarrow Fe°$$

For organic molecules, oxidation often involves loss of hydrogen atoms by reaction with oxygen and reduction involves a gain of hydrogen atoms. The browning of a freshly cut apple or banana is a common example of organic oxidation and the manufacture of oleomargarine from vegetable oil makes use of organic reduction.

In the present experiment we will reduce the blue, water-insoluble dye indigo to a water-soluble, colorless form (leuco-indigo), then heat a piece of cloth in the solution, and then air oxidize the leuco-indigo that is adhering to the cloth back to the deep blue water-insoluble indigo.

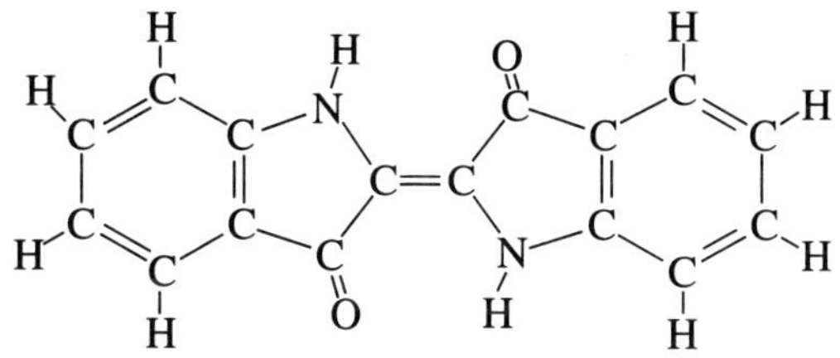

Indigo

Oxygen ⇅ $Na_2S_2O_4$

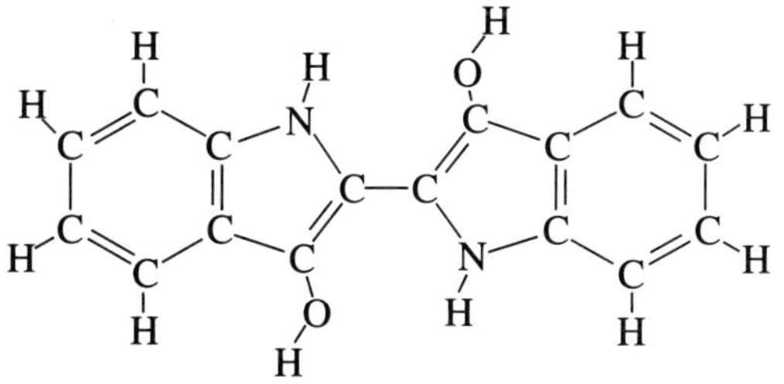

Leuco indigo, the reduced form

Indigo is not like any other dye used to color your clothes because it can be rubbed off the surface of the fiber on which it is deposited. Most dyes form chemical bonds—covalent bonds—between the dye and the fiber or else they dissolve inside the fibers. Leuco-indigo is attracted to cotton (in particular) by much weaker hydrogen bonds. On exposure to the oxygen in air the leuco form is oxidized back to the dye, which simply coats the surface of the fiber. If the fiber is abraded in places of wear, for example, the knees of jeans, the dye comes off and the cloth appears whiter. Or the entire garment can be tumbled with pumice rocks (stone-washed jeans) to give a very characteristic appearance.

Procedure Summary

Water-insoluble blue indigo is reduced to a colorless water-soluble leuco form. Cloth is heated in this solution; the leuco dye adheres to the cloth. On exposure to air the reduced form of the dye is oxidized back to water-insoluble blue indigo.

Prelaboratory Assignment

Read the Introduction and Procedure sections carefully noting especially how much sodium hydrosulfite is required. Answer the Prelaboratory Question on the Report Sheet.

Materials

Apparatus

Electrically heated sand bath or hot plate
Thermometer (optional)
30-mL beaker
Three-prong clamp or small ring to support beaker
Glass rod or tongs
Ring stand
Metal spatula
Paper towels
Cotton fabric (untreated) or multifiber cloth, 2-inch squares

Reagents

3 M sodium hydroxide
Indigo
Sodium hydrosulfite
Liquid or solid soap or detergent
3% hydrogen peroxide (optional)
Household bleach (5.25% NaOCl) (optional)

Experiment 41 Oxidation–Reduction: Dyeing with Indigo, the Blue Jeans Dye

> **Safety Information:**
> 1. **Safety goggles must be worn at all times in the laboratory.**
> 2. **Wipe up spills immediately.** Anchor the beaker of hot water so that it does not spill.
> 3. **Avoid contact with sodium hydroxide solution and sodium hydrosulfite and the leuco-indigo solution.** If any of these get on the skin, wash them off with plenty of water.

Procedure

1. To a 50-mL beaker containing about 30 mL of water add about 50 mg of indigo. Stir the mixture. Usually the indigo will float on the surface of the water. It is not wetted by the water. Add a very small drop of liquid detergent or a pinch of soap powder to the beaker. This will reduce the surface tension of the water. When the mixture is stirred the indigo will disperse in the water giving it a blue appearance. But the dye is present as tiny particles; it does not dissolve in the water and could be removed by filtration. Put a drop of this suspension on a piece of filter paper and observe the results.

2. Heat the aqueous indigo dispersion to about 80°C and, while stirring, add to it about 1.3 mL of 3 M sodium hydroxide solution followed by about 0.25 g of sodium hydrosulfite. The indigo should react to give a transparent solution that has a yellowish color. You should be able to see through the solution from the side of the beaker. From the top the surface will appear blue. If all of the indigo does not react, add more sodium hydrosulfite.

3. Add a 2 × 2-inch piece of cotton or a piece of multifiber cloth to the beaker and heat it for 5 minutes. Remove the cloth with a glass rod or a pair of tongs and wash it immediately with soap under running water or in a container of soapy water. Rinse the cloth and pat it dry with a paper towel. Note carefully the color of the cloth while it is being dyed in the hot leuco-indigo solution, while it is being washed, and then over the next 15 or 20 minutes. Put a drop of the leuco-indigo solution on a piece of filter paper and observe the results.

4. Perform the following tests on pieces cut from your dry, dyed cloth sample:
   a. Place a small piece of your cloth sample in about 2–3 mL of 3% H_2O_2 (household antiseptic). Describe and observe the results.
   b. Place a small piece of your cloth sample in about 2–3 mL of ordinary liquid bleach, 5% NaOCl(aq). Describe and observe the results.

Cleaning Up

Pour the indigo dye bath into a large open container and allow it to oxidize back to water-insoluble indigo. Neutralize the mixture with dilute acid and either flush it down the drain (the amount of dye is quite small) or filter it off and dispose of it in the nonhazardous solid waste container. The filtrate can be washed down the drain.

Name _____   Section _____

Lab Instructor _____   Date _____

EXPERIMENT 41 Oxidation–Reduction: Dyeing with Indigo

PRELABORATORY QUESTION

1. What is the purpose of soap in this experiment? Of sodium hydrosulfite?

DATA, OBSERVATIONS, AND CONCLUSIONS

1. What is the appearance of the reaction mixture immediately after the indigo is added to the beaker of water?

2. What is the effect of adding a drop of detergent or soap solution to the beaker of water and stirring?

3. Describe the appearance of the filter paper on which a drop of the indigo suspension was placed. Was the indigo soluble in the water?

4. What is the appearance of the reaction mixture after sodium hydrosulfite has been added to it?

5. Why is the solution blue when observed from above but light yellow when observed from the side?

6. Describe the appearance of the filter paper on which a drop of the leuco-indigo solution was placed (a) immediately (b) after the lapse of a period of time. How do you explain these observations? Is leuco-indigo in solution?

7. Describe the appearance of the cloth immediately after removing it from the dye bath and washing it with soap and water.

8. Describe the appearance of the cloth after 1, 5, 10 and 20 minutes. How do you account for these changes?

9. What would you expect to happen if you added a piece of indigo-dyed cloth to a hot solution of alkaline sodium hydrosulfite? Explain your reasoning.

10. Hydrogen peroxide is an oxidizing agent. Describe and explain your observations in Step 4a.

11. Describe and explain your observations in Step 4b. Recall earlier experiments in which bleach (5.25% sodium hypochlorite solution) was used as an oxidizer.

12. Examine the chemical formula for indigo and leuco-indigo. Does each hydrogen, oxygen, nitrogen and carbon have a valence of 1, 2, 3, and 4, respectively?

13. During the reduction process, which two hydrogens seem to have been added to the indigo?

14. Look up the definition of a hydrogen bond, and also look up the chemical structures of wool (protein), cotton (cellulose), and polypropylene $(CH_2CH(CH_3))_n$. Explain why some fibers dye much better than others. (You saw these effects if you dyed a multifiber fabric sample.)

POSTLABORATORY QUESTIONS

1. Dyeing with indigo has been done for many thousands of years. Speculate upon or look up the process whereby the reduction was carried out before the advent of a chemical reducing agent such as sodium hydrosulfite.

2. Indigo is unique among the common dyes in that it can be rubbed from the fiber, hence the unique appearance of "stone-washed" jeans. Speculate why other dyes do not rub off the fabric.

EXPERIMENT 42

Synthesis of Slime

Introduction

Undoubtedly you have encountered "slime," perhaps in a grade school science demonstration. It is fun to play with, but it also has much to teach us about polymer chemistry, charge distributions, the nature of hydrogen bonds, the chemistry of boron (and silicon) and the hydroxyl group, polymer cross-linking, polymer chains, and three-dimensional structures.

The commercial product known as Slime™ is produced by the Mattel Toy Corporation and is slightly different from the material we will synthesize. Our slime is produced by mixing a 4% solution of poly(vinyl alcohol) with boric acid or sodium tetraborate decahydrate (borax). The result is a slimy semisolid gel-like substance that can be picked up, stretched, broken, molded, and bounced even though it is 96% water. The object of this experiment is to investigate the forces that hold this slime together.

Poly(vinyl alcohol) is a long chain of carbon atoms with hydroxyl groups on every other carbon atom. It is a linear polymer and can have a molecular weight of up to 150,000 with almost 7,000 carbon atoms in a row. Ordinarily such a high-molecular-weight substance would not be soluble in water, but the large number of hydroxyl groups that can form hydrogen bonds with water contribute strongly to this remarkable solubility.

Poly(vinyl alcohol)

Boric acid is a very weak acid that undergoes hydrolysis in water where it forms a boric acid/borate buffer system:

$$B(OH)_3 + 2\,H_2O \rightleftharpoons B(OH)_4^- + H_3O^+ \qquad pK_a = 9.2$$

Boric acid is such a weak acid that it does not give up a proton:

but instead accepts an OH⁻ from the water

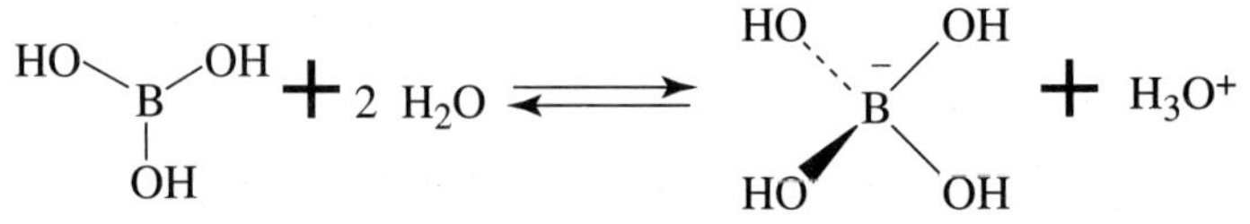

$$\text{HO-B(OH)}_2\text{-OH} + 2\ H_2O \rightleftharpoons [\text{B(OH)}_4]^- + H_3O^+$$

Coplanar   Tetrahedral

There is the possibility that boric acid or the tetrahedral borate ion could form borate esters with poly(vinyl alcohol) to give a three-dimensional network in which chains of the polyalcohol are bound to each other or even to other parts of the same chain. Using the borate ion as an example:

[Diagram showing two poly(vinyl alcohol) chain segments with OH groups plus borate ion $[B(OH)_4]^-$ reacting to form a cross-linked structure with B bonded to four O atoms from the polymer chains, releasing $4\ H_2O$]

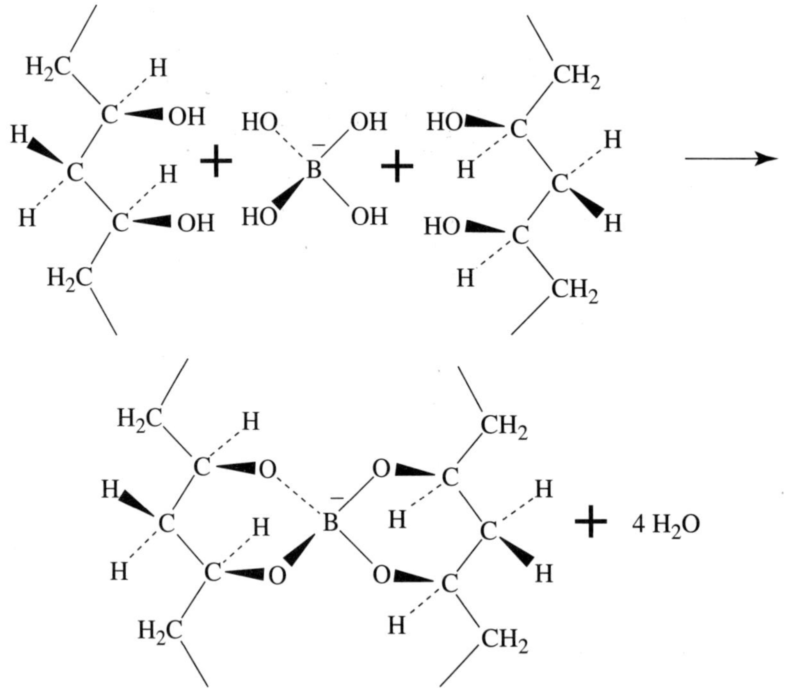

But, as logical as this picture may seem, it is probably not correct. If covalent bonds were formed between the borate and the polymer chain, we would not expect the resulting cross-linked polymer to dissolve in water, as slime does. In addition, the physical properties of slime—the way it flows and can be stretched and bounced—are not compatible with this picture. In general, cross-linked organic polymers do not dissolve in any solvent and once cross-linked cannot be melted or reformed. Bakelite, from which the black knobs and handles on cooking utensils are made, is an example of a highly cross-linked polymer.

Another picture of slime would entail the formation of hydrogen bonds between the oxygen atoms of boric acid or borate ion and the hydroxyl hydrogens of the poly(vinyl alcohol). Again using the borate ion as an example:

[Structural diagram showing poly(vinyl alcohol) chains cross-linked by borate ion through hydrogen bonds]

Presumably boric acid could do exactly the same thing:

[Structural diagram showing poly(vinyl alcohol) chains cross-linked by boric acid B(OH)₃ through hydrogen bonds]

These hydrogen bonds (represented by the dotted lines) would give the network stability, but when subjected to a shearing stress they could be broken and then reform elsewhere. This picture would explain why the polymer is so readily soluble in a large excess of water and why it tends to turn fluid under shearing stress.

Bear in mind that hydrogen bonds result when the hydrogen donor atom carries a partial positive charge and the receptor oxygen atom in the other molecule carries a partial negative charge and a pair of unshared electrons.

Hydrogen bonds are broken as the temperature of the system rises. What would be the expected effect of heating slime? When you handle slime try to picture, on a molecular scale, exactly what you are holding. In particularly think about the percentage of water in slime.

It is instructive to compare boric acid with silicic acid, H_4SiO_4. Note the relative positions of boron, carbon, and silicon in the periodic table. Unlike boric acid, silicic acid is unstable and in dilute acid forms siloxane bonds, O–Si–O, to give a three-dimensional network, a gel. In base it does not do this and is, in fact, capable of forming hydrogen bonds with poly(vinyl alcohol) in much the same way that boric acid does. A variant on slime can be made using sodium silicate.

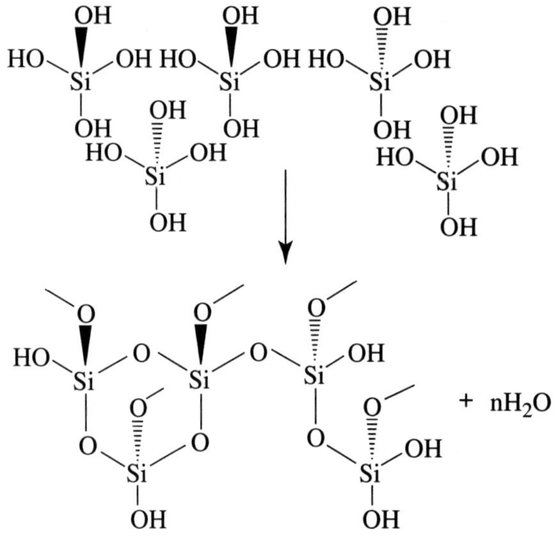

Experiment 42   Synthesis of Slime

Procedure Summary

Slime is very easily prepared by simply stirring a solution of boric acid and base or sodium borate into a solution of poly(vinyl alcohol). In a minute or two it can become quite thick and in about 10 minutes it can set to a firm gel if the proper conditions are met.

Prelaboratory Assignment

Read the Information and Procedure sections carefully and answer the Prelaboratory Questions on the Report Sheet.

Materials

Apparatus

Paper cups, small
Wood applicator sticks
10-mL graduated cylinder
1-mL graduated pipette
pH paper or Universal Indicator

Reagents

4% Weight/volume aqueous poly(vinyl alcohol) solution
0.2 M aqueous boric acid solution
0.2 M aqueous sodium hydroxide solution
0.2 M aqueous silicic acid solution

Safety Information

1. **Safety goggles must be worn at all times in the laboratory.**
2. **Slime prepared from sodium tetraborate is not toxic, nor is it corrosive.** It washes out of clothes, but is difficult to remove from rugs and carpets.
3. **The substance made with sodium silicate is very basic.** Handle with care.

Procedure

1. Measure the pH of 1 mL of 0.2 M boric acid solution and of the solutions that arise when 1, 2, and 3 mL of 0.2 M sodium hydroxide solution are added to the boric acid.
2. In a paper cup place 10 mL of 4% poly(vinyl alcohol) solution. To this add 2 mL of 0.2 M boric acid solution. Stir the mixture thoroughly. A wood boiling stick (applicator stick) works very well for this. Note the appearance and viscosity of the mixture.
3. Add 1 mL of 0.2 M sodium hydroxide solution. Stir the mixture well and observe its viscosity.
4. Add 1 mL more of 0.2 M sodium hydroxide solution. Stir the mixture well and again note the nature of the mixture.
5. To a fresh cup add 10 mL of 4% poly(vinyl alcohol) solution. To this add a mixture of 2 mL of 0.2 M boric acid and 3 mL of 0.2 M sodium hydroxide solution. Stir the mixture thoroughly and note the nature of the mixture.
6. Remove the material from the first cup, place it in a small beaker and heat it on a steam bath or hot plate for about 5 minutes with stirring.
   a. What change in physical properties is noted in the hot mixture? After cooling to room temperature the mixture can be removed from the beaker. Roll the slime into a ball and place on a flat surface.
   b. What happens to it over a period of time?
   c. Hit the relaxed ball of slime hard with your hand. What happens to it?
   d. Pick it up and stretch it slowly. Note what happens. Try to determine to what length you can stretch it without having it break.
   e. Roll it into a cylinder and pull on it rapidly. Note what happens.
   f. Can you bounce a ball of slime?
7. Try dissolving a small piece of slime in about 25 mL of water. Stir the mixture thoroughly.

8. In a third cup place 2 mL of 0.2 M boric acid solution and 2 mL of 0.2 M sodium hydroxide solution. To this mixture add about 1 g of sodium chloride and stir the mixture thoroughly to dissolve as much of the salt as possible. Add 10 mL of 4% poly(vinyl alcohol) solution and stir the mixture. Compare the results with the material from Steps 3 and 6.

Sodium Silicate Experiments

9. To a fourth paper cup add 2 mL of sodium silicate solution and 10 mL of 4% poly(vinyl alcohol) solution. Stir the mixture for about 1 minute and collect the product on the end of the wood applicator stick. Rinse it briefly with water in a beaker and then squeeze out excess water between paper towels. Compare the properties of this material with that made earlier. You might also try mixing equal volumes of sodium silicate solution with the poly(vinyl alcohol) solution. Note the properties of this material a day or so after it has been prepared.

Cleaning Up

The rigid and semirigid polymers can be disposed of in the nonhazardous waste container. Liquid solutions can be washed down the drain with much water.

EXPERIMENT 42   Synthesis of Slime

PRELABORATORY QUESTIONS

1. Based on the discussion of how gels form and are dependent on temperature and pH, what would you predict for the outcome of Steps 2 through 4? Which would you expect to be the most viscous and slime-like? Explain. What would you predict would be the effect of heating slime? Will it become more or less solid?

2. Which is the stronger acid: H_2O or H_3BO_3? Which is the stronger base: H_2O or H_2BO_3? Identify the conjugate acid/base pairs in the equation given in the Introduction for the reaction between boric acid and water.

DATA

1. pH of 0.2 M boric acid   _____

    pH of 0.2 M boric acid plus 1 mL 0.2 M NaOH   _____

    pH of 0.2 M boric acid plus 2 mL 0.2 M NaOH   _____

    pH of 0.2 M boric acid plus 3 mL 0.2 M NaOH   _____

OBSERVATIONS AND CONCLUSIONS

2. Boric acid + poly(vinyl alcohol) observations

3. Boric acid + poly(vinyl alcohol) + 1 mL NaOH observations

4. Boric acid + poly(vinyl alcohol) + 2 mL NaOH observations

5. Boric acid + poly(vinyl alcohol) + 3 mL NaOH

6. a. Change on heating

   b. Change of ball over time

   c. Hitting slime

   d. Slowly stretching slime

e. Rapidly stretching slime

f. Bouncing slime

7. Dissolving slime in water

8. Addition of sodium chloride

9. Slime made with sodium silicate

POSTLABORATORY QUESTION

1. How would you summarize your observations in this experiment? Write a short summary in complete sentences.

APPENDIX

Useful Data Tables

Table 1 The International System (SI) of Units and Conversion Factors

Basic SI units

Physical Quantity	Unit	Symbol
Length	meter	m
Mass	kilogram	kg
Time	second	s
Temperature	kelvin	K
Amount of substance	mole	mol

Common Derived Units

Physical Quantity	Unit	Symbol	Definition
Energy	joule	J	$kg \cdot m^2 \cdot s^{-2}$
Force	newton	N	$kg \cdot m \cdot s^{-2} = J \cdot m^{-1}$
Pressure	pascal	Pa	$kg \cdot m^{-1} \cdot ^{-2} = N \cdot m^{-2}$

Decimal Fractions and Multiples

Factor	Prefix	Symbol
10^{-12}	pico	p
10^{-9}	nano	n
10^{-6}	micro	μ
10^{-3}	milli	m
10^{-2}	centi	c
10^{-1}	deci	d
10	deca	da
10^2	hecto	h
10^3	kilo	k
10^6	mega	M
10^9	giga	G

Common Conversion Factors

LENGTH
1 angstrom unit (Å) = 10^{-8} cm
2.54 cm – 1 inch (in.)
1 m = 39.4 in.

MASS
453.5 grams (g) = 1 pound (lb)
1 kg = 2.20 lb
28.3 g = 1 ounce (oz, avoirdupois)

VOLUME
1 milliliter (mL) = 1 cubic centimeter (cm^3)
 (Note that the mL is now defined precisely equal to 1 cm^3.)
1 liter (L) = 1.06 quarts
28.6 mL = 1 fluid ounce

PRESSURE
1 atm = 1.013×10^5 Pa (N/m^2) = 101.3 K Pa
 = 760 torr (mmHg); pressure of a mercury column 760 mm or 29.92 in. high at 0°C
 = 14.70 lb/in.^{2}

TEMPERATURE
Absolute zero (K) = –273.15°C
K = °C + 273.15
Fahrenheit degrees (°F) = $\frac{9}{5}$°C + 32
Celsius degrees (°C) = $\frac{5}{9}$ (°F – 32)

ENERGY
1 joule = 1 W-s = 10^7 erg
1 erg = 1 dyne-cm = 1 $g \cdot cm^2 \cdot s^{-2}$
1 calorie = 4.184 (J)
1 electron volt/molecule = 23.06 kcal/mol

Table 2 Fundamental Physical and Mathematical Constants

Physical Constants

Symbol	Name	Numerical Value
N_A	Avogadro's number	6.0221×10^{23} mol^{-1}
R	The gas constant	0.08206 L·atm·mol^{-1}·K^{-1}
		82.06 mL·atm·mol^{-1}·K^{-1}
		8.314 J·mol^{-1}·K^{-1}
	Volume of 1 mol of an ideal gas	
	at 1 atm 0°C	22.41 L
	at 1 atm 25°C	24.46 L

Mathematical Constants

$\pi = 3.1416$

$\ln x = 2.303 \log_{10} x$

Table 3 Acids and Bases

	Specific Gravity	% by Weight	Moles per Liter	Grams per 100 mL
Hydrochloric acid, concd	1.19	37	12.0	44.0
Constant boiling (252 mL concd. acid + 200 mL water, BP 110°)	1.10	22.2	6.1	22.2
1 M (41.5 mL concd. acid diluted to 500 mL)	1.02	3.6	1	3.6
Sulfuric acid, concd.	1.84	96	18	177
1 M (13.9 mL concd. acid diluted to 500 mL)	1.03	4.7	0.5	4.9
Nitric acid, concd.	1.42	71	16	101
Sodium hydroxide, 10% solution	1.11	10	2.8	11.1
Ammonia solution, concd.	0.90	28.4	15	25.6

concd. = concentrated.

Table 4 Buffer Solutions (0.2 M, except as indicated)

pH	Components	pH	Components
0.1	1 M Hydrochloric acid	8.0	11.8 g Boric acid + 9.1 g Borax
1.1	0.1 M Hydrochloric acid		(Na$_2$B$_4$O$_7$·10H$_2$O) per L
2.2	15.0 g D-Tartaric acid per L (0.1 M solution)	9.0	6.2 g Boric acid + 38.1 g Borax per L
3.9	40.8 g Potassium acid phthalate per L	10.0	6.5 g NaHCO$_3$ + 13.2 g Na$_2$CO$_3$ per L
5.0	14.0 g KH-Phthalate + 2.7 g NaHCO$_3$ per L	11.0	11.4 g Na$_2$HPO$_4$ + 19.7 g Na$_3$PO$_4$ per L
	(heat to expel carbon dioxide, then cool)	12.0	24.6 g Na$_3$PO$_4$ per L (0.15 M solution)
6.0	23.2 g KH$_2$PO$_4$ + 4.3 g Na$_2$HPO$_4$	13.0	4.1 g Sodium hydroxide pellets per L (0.1 M)
	(anhyd., Merck) per L	14.0	41.3 g Sodium hydroxide pellets per L (1 M)
7.0	9.1 g KH$_2$PO$_4$ + 18.9 g Na$_2$HPO$_4$ per L		

Table 5 Vapor Pressure of Water at Different Temperatures

Temperature (°C)	Vapor Pressure (mm Hg)
−10 (ice)	1.0
−5 (ice)	3.0
0	4.6
5	6.5
10	9.2
15	12.8
16	13.6
17	14.5
18	15.5
19	16.5
20	17.5
21	18.6
22	19.8
23	21.1
24	22.4
25	23.8
26	25.2
27	26.7
28	28.3
29	30.0
30	31.8
35	42.2
40	55.3
45	71.9
50	92.5
55	118.0
60	149.4
65	187.5
70	233.7
75	289.1
80	355.1
90	525.8
100	760.0
150	3570.5
200	11659.2